SOIL DEGRADATION AND ITS IMPACT ON THE NUTRIENT CONTENT OF FOOD

SOIL DEGRADATION AND ITS IMPACT ON THE NUTRIENT CONTENT OF FOOD

Dr. Sara D. Garduño-Diaz

Delve
PUBLISHING

www.delvepublishing.com

Soil Degradation and its Impact on the Nutrient Content of Food

Dr. Sara D. Garduño-Diaz

Delve Publishing
4164 Lakeshore Road
Burlington, ON L7L 1A4
Canada
www.delvepublishing.com
Email: orders@arclereducation.com

ISBN: 978-1-77956-467-2 (Hardcover)

ABOUT THE AUTHOR

Sara D. Garduño-Diaz holds a PhD in Nutrition and Food Science from the University of Leeds, United Kingdom, and has further specialized in Sports Nutrition with the International Olympics Committee, Switzerland. Dr. Garduño-Diaz has worked as a Research Fellow for several EU-funded research projects and has been awarded research grants from CONACYT (Mexico) and the BBSRC (UK). She currently consults for various startups in Europe and the US in the areas of nutrition and product development.

Dr. Garduño-Diaz sits on the editorial board of several international journals, including Journal of Obesity and Weight Management, E Cronicon Nutrition, Research in Health Science, and Insights in Nutrition and Dietetics. She is the author of over 50 peer-reviewed publications and 10 books. Her research interests include the dietary patterns of migrant populations and their impact on health, as well as the environmental factors influencing food selection.

Dr. Garduño-Diaz is a member of the Academy of Nutrition and Dietetics, the American Overseas Dietetics Association, Professionals in Nutrition for Exercise & Sport, and the World Public Health Nutrition Association, where she serves on the executive committee.

With over 15 years of experience in the field of nutrition and having lived and worked across various continents, Dr. Garduño-Diaz's approach to food integrates strategies to design individual programs based on her clients' wellness needs. She advocates for real food and the art of eating.

TABLE OF CONTENTS

LIST OF FIGURES

LIST OF TABLES

PREFACE

The historical uneasiness with which scientists have embraced the concept of soil health stems from the challenges of defining soil health in a way that allows for a universal quantitative assessment encompassing all of its ecosystem services, including human health. Reasons for this challenge include soil heterogeneity, the site-specific nature of soil management, and the varying ecosystem services, which sometimes have conflicting or competing needs. Nevertheless, there has been widespread interest among researchers, policymakers, and stakeholders in the use of the soil health concept.

There are many components in the connection between nutrition and farming, and much remains to be fully understood. As is often the case, more research is needed; however, with more than two billion people worldwide impacted by micronutrient deficiencies, significant benefits could arise from investigating the trail of nutrients.

The most widely recognized function of soil is its support for food production. It is the foundation of agriculture and the medium in which nearly all food-producing plants grow. It is estimated that approximately 95% of food worldwide is directly or indirectly produced on soils. Healthy soils supply the essential nutrients, water, oxygen, and root support that food-producing plants need to grow and flourish. Soils also serve as a buffer to protect delicate plant roots from drastic fluctuations in temperature.

Soil quality plays a crucial role in the production of nutrient-rich food. The health of the soil directly impacts the nutrient content of crops, making it essential for food scientists, university students, researchers, and policymakers—along with farmers and other food producers—to understand this connection. In this book, the concept of soil health and nutrient quality in food is explored.

–Author

CHAPTER 1

GLOBAL HUNGER

Contents

The international hunger crisis is a significant on-going challenge, with its origins varying from nation to nation, particularly in areas worst hit by military or security conflicts (WFP, 2022). Countries such as Afghanistan, Ethiopia, and Yemen are currently undergoing concerning levels of hunger (30.6, 26.2, and 39.9%, respectively), while Somalia and South Sudan are temporarily under the category of alarming (35.0–49.9%) (GHI, 2023). Additionally, Nigeria also has a level of hunger that is alarming with 28.3% of its population experiencing hunger, ranking 109th out of the 125 countries considered as part of the Global Hunger Index (GHI, 2023). The global hunger index score is categorized as low (≤9.9), moderate (10.0–19.9), serious (20.0–34.9), alarming (35.0–49.9), and extremely alarming (≥50.0).

From a World point of view, previous to the onset of the COVID-19 pandemic, approximately 820 million people had been affected by significant levels of hunger and starvation (Kakaei et al., 2022), with approximately 282 million people in Africa being undernourished (Mohajan, 2022), of which 220 million are from sub-Saharan Africa (Obada et al., 2021). This is in part due to African countries lacking the prioritization of food production and failing to address the reduction of food insecurity, hunger, and malnutrition. Certainly, as it has been pointed out in the past, ending hunger would be a disaster for some people, as hunger is not a problem but an asset to those at the top end of the social ladder (Kent, 2008).

1.1. HUNGER AND FOOD INSECURITY

Food insecurity is the official term referring to the occurrence of people not having enough to eat and not knowing where their next meal will come from. Food insecurity is a significant problem globally and it affects not only developing, but also developed countries. For example, in the United States, it is estimated that over 44 million people, including 13 million children, experience food insecurity every year. Food insecurity is a systemic challenge that can affect anyone and must not be considered as a personal failure. According to Feeding America's Elevating Voices Report, people facing hunger struggle with high living costs, expensive housing, unemployment, and low-paying employment. Hunger and food insecurity are related challenges that cover inadequate access to nutritious, safe and sufficient food. Food insecurity can thus range from a mild and short-lived shortage of food to hunger, meaning a severe and chronic lack of food. In a world where sufficient food for everyone is produced, millions remain hungry and experience food insecurity.

Understanding the root causes of the global food crisis is paramount to finding effective solutions. For example, reducing food loss and waste could in turn render enough food available to feed malnourished people worldwide.

Managing food resources in a more effective manner could alleviate hunger and improve food security, ensuring that food reaches those who need it most.

The World Bank estimates that more than 648 million people live in extreme poverty, below the international poverty line of $2.15 per day. Poverty limits people's ability to afford food and their access to education and healthcare, perpetuating the cycle of poverty and food insecurity. The Food and Agriculture Organization of the United Nations states that "climate change poses a major and growing threat to global food security." Climate change also impacts the livelihoods of farmers and the health of vulnerable communities.

Although some regions of the World achieve an excess of food production, others face on-going shortages. According to The Global Hunger Index, Africa and South Asia have the highest levels of hunger in part due to restrained access to technology and infrastructure to set up a food distribution system that is effective.

The impact of hunger around the world is deep and multifactorial. Hunger is an international threat that can cause severe health issues. Kids, particularly those in the first 1,000 days of life, are at risk of malnutrition, developmental delay, impaired growth, and possibly infant and child mortality. Women who are pregnant can experience complications, including miscarriages, premature delivery and maternal death. Hunger can damage the immune system, resulting in illness and disease vulnerability, which can be fatal to the elderly or those with chronic illnesses.

Mental health and well-being can also be affected when experiencing hunger and food insecurity. Food insecurity may lead to an increase in stress, hypertension, depression, anxiety and conflict. Lack of quality nutrition can impair the ability to learn and future potential, continuing the cycle.

1.1.1. Terminology Around the Concept of Hunger and Food Insecurity

Food Insecurity refers to the inability to consistently meet nutritional and caloric needs. Households that are food insecure may be concerned with running out of food, may not meet suitable diversity of nutrients or may compromise in the quality and number of meals.

Hunger is the physical pain caused by not enough food consumption. Regarding humanitarian aid and global development, hunger is much more concerning, referring to the experience of ongoing periods of time without consuming enough calories to maintain health. Chronic hunger often ends in malnutrition and even starvation, turning into a strong threat for mental and physical health and child development.

- **Poverty:** Poverty is a main cause of food insecurity. Impoverished communities often do not have access to resources, education and job opportunities, making it difficult for them to secure sufficient food for their families.
- **Climate Change:** Climate change impacts agriculture, limiting crop yields and reducing the availability and affordability of food. Extreme weather events, including droughts and floods increase food insecurity.
- **Inequity:** Distribution of natural and economic resources that is unequal can perpetuate hunger. Many communities have insufficient access to nutritious food because of systemic inequities.
- **Food Loss and Waste:** Approximately 30% of all food produced globally is never eaten because of food loss and waste. This loss to the global food supply translates to 1.3 billion tons of food that could have potentially fed those in need.
- **Conflict:** Conflict contributes strongly to food insecurity. When violence takes place, agricultural production is often affected, infrastructure can be disrupted and people can be displaced, making it difficult to have access to food.

Box 1: Food Waste

Food waste is formed along the full spectrum of production, from the farm to distribution to retailers to the consumer. The causes for food waste vary and may include losses from mold, pests, or inadequate climate control; losses as part of the cooking process, as well as intentional food waste (USDA, 2017). This waste is categorized differently based on where it occurs:

- Food "loss" takes place before the food reaches the consumer due to issues with the production, storage, processing, and distribution phases of the food in case.
- Food "waste" in turn happens to food that is fit for consumption but is consciously discarded at the retail (market) or consumption (home, restaurant) phases.

Food waste has far-reaching effects internationally. For example, in the USA, up to 40% of all food produced goes uneaten (Gunders et al., 2012), and approximately 95% of discarded food ends up in landfills (USEPA, 2017). In the USA, food waste is the largest component of municipal solid waste at 21% (USDA, 2017).

In 2014 alone, over 38 million tons of food waste was created, with a mere 5% moved from landfills and incinerators for composting (USEPA, 2017). In turn, the decomposition of food waste produces methane, a strong greenhouse gas that contributes to global warming. Globally, one-third of food produced is thrown away uneaten, causing a higher burden on the environment (Salendeeb et al., 2017). It is estimated that lowering food waste by 15% could feed more than 25 million people annually (Hall et al., 2009).

1.2. ADDRESSING HUNGER AND FOOD INSECURITY REQUIRES A MULTIFACETED APPROACH

Actions to tackle world hunger and food insecurity require a myriad of approaches, including bettering food production and distribution systems, championing sustainable agriculture, lowering poverty and inequity, strengthening social safety nets and addressing the root causes of these issues, including conflict and climate change.

International associations, governments, nonprofits and individuals all play key roles in moving towards a global setup where all have access to suitable and nutritious food, ultimately targeting the United Nations Sustainable Development Goal of zero hunger.

The championing of sustainable agriculture practices that allow various demographic books to produce their own food while looking after the Earth is a single strategy to improving the state of food insecurity. By facilitating training and giving tools to micro-scale farmers, it is possible to increase crop and animal yields and improve income while considering the maintenance and restoration of the environment.

Extreme climate conditions and changing seasonality impacts the ability of farmers to nourish their families and their communities, leading to increases in food insecurity and hunger. Using practices from climate-smart solutions, micro-scale farmers pick up new practices and technologies that create resilience and protect limited natural resources, including soil and water.

An alternative likely favorable practice is the championing of regenerative agriculture, a type of farming that replenishes the soil and augments biodiversity. From utilizing these practices it is possible to train micro-scale farmers to manage their land in a holistic manner to improve productivity and develop more resilient food systems.

Box 2: Global Hunger Index

The Global Hunger Index (GHI) is a peer-reviewed report published every year, between the groups of Concern Worldwide and Welthungerhilfe, created to robustly determine the degree of hunger at the international, regional, and national levels. The objective of the GHI is to set off activities to lower hunger globally.

The global hunger index considers three dimensions of hunger: lacking food availability, inadequacy in the nutritional status of kids and kid mortality (connected, to a great extent, with a state of undernutrition). Thus, the index is made up of three equally contributing indicators: the proportion of individuals who are food energy-deficient, as estimated by Food and Agriculture Organization; the prevalence of underweight in kids aged below 5 years, as compiled by World Health Organization; and the mortality rate of kids aged below 5 years, as reported by UNICEF. A regression analysis of the global hunger index on GNI per capita is carried out to pinpoint nations that are significantly better or worse off in terms of hunger and undernutrition than would be expected from their GNI per capita.

Hunger is one of the world's most pressing challenges and, for that reason, one of its most important problems to be addressed. Hunger and undernourishment form a vicious circle, which is regularly "passed on" from one generation to the next. The children of poor parents are regularly born underweight and are more vulnerable to disease; these kids grow up under conditions that hinder their intellectual capacity for the duration of their lives. Issues that contribute to a high global hunger index have been pinpointed as low income and poverty, war and violence, general lack of freedom, women's status as low, and poorly targeted and delivered health and nutrition programs.

Source: WHO (2024).

CHAPTER 2

THE IMPORTANCE OF SOIL QUALITY IN FOOD PRODUCTION

Contents

The most widely recognized function of soil is its support for food production. It is the foundation for agriculture and the medium in which nearly all food-producing plants grow. It is estimated that approximately 95% of food worldwide is directly or indirectly produced on soils. Healthy soils supply the essential nutrients, water, oxygen and root support that food-producing plants need to grow and flourish. Soils also serve as a buffer to protect delicate plant roots from drastic fluctuations in temperature.

Soil quality plays a crucial role in the production of nutrient-rich food. The health of the soil directly impacts the nutrient content of crops, making it essential for food scientists, university students, researchers, and policy makers to understand this connection. This of course, in addition to the farmers and other food producers directly.

One key factor in soil quality is its pH levels. The acidity or alkalinity of the soil can have a significant impact on the availability of nutrients to plants. Soil with the proper pH balance can help ensure that plants are able to take up essential nutrients, resulting in healthier and more nutrient-dense food.

Organic farming practices are also important in maintaining soil quality. Compared to conventional farming methods, organic practices focus on building soil health through the use of compost and other natural amendments. This helps improve soil structure and fertility, leading to higher nutrient content in food crops (Brinton, 2015).

Soil erosion is another concern that can affect nutrient levels in crops. Erosion can deplete the soil of essential nutrients, making it crucial to implement sustainable agriculture practices to prevent this loss. By maintaining soil health, farmers can ensure that their crops are able to access the nutrients they need to thrive (Altieri, 2018).

Soil microbiology also plays a vital role in nutrient availability for plants. The microbial community in the soil helps break down organic matter and release nutrients for plant uptake. Understanding the role of soil microbes can help farmers make informed decisions about soil amendments and fertilizers to improve nutrient content in food.

Overall, the importance of soil quality in food production cannot be overstated. By prioritizing soil health and implementing sustainable practices, we can ensure that our food is not only abundant but also rich in essential nutrients for optimal health.

In the most basic sense, soils are the basis for life on the Earth. Soils are the support medium that plants rely on to remain standing. Soils provide the habitat for ample biodiversity and biomass of soil organisms (Nielsen et al., 2015), and both create and function as the repository for most of the carbon and nutrient elements that support life (Brevik and Sauer, 2015). Soils help with the retention of the water that plants and soil organisms utilize to survive and grow, and reduce the rate of water movement and thus cap the rate of erosion and soil loss (Horton, 1933). Additionally, soils add to the atmospheric composition, and thus have an impact on climate, and are both a significant source and sink of greenhouse gasses (Oertel et al., 2016; Mayer et al., 2018).

At the level of society, one of the most obvious contributions of soil to people is the role that soils have in the provision of food for human populations and feed for livestock. People have directly managed soils for food and feed production via agriculture for more than 12 thousand years (Zeder, 2011). For over a million years, soils have been indirectly utilized for food and feed production through hunting and gathering (Mercader, 2009). About 40% of the terrestrial land surface is currently dedicated to food and feed production, with an estimated 12% in crop agriculture and 25% in grazing lands (Diaz et al., 2018).

The activities related to food and feed production in many ways resemble those of mining. Plants pull nutrients from soils and keep them in their tissues. Plants are in turn harvested and thus transport soil nutrients out of the ecosystem, a process that is only one step removed when livestock convert plants to animal biomass that itself is raised for human consumption. Nutrients then return to soils in the form of organic (plant, livestock and other organic matter resources) or inorganic fertilizers and are critical to the upkeep of soil fertility in agricultural soils. Nevertheless, nutrient inputs seldom fully replace the nutrients extracted via harvest, thus leading to soil nutrient depletion and degradation over the long term (Sanchez, 2002; Vitousek, 2009; Tan et al., 2005).

Soils vary significantly in the availability of the resources required for food and feed production. As an example, old, highly weathered soils tend to contain low phosphorus (P) availability in part due to the high P sorption capacity of residual iron and aluminum minerals, linked with deep profiles and the lack of new P inputs from weathering products of primary minerals (Vitousek et al., 2010; Walker et al., 1976). Inherent nitrogen (N) limitation is usual in recently unglaciated regions and new, volcanically-derived soils, as well as in high latitude regions with low mean yearly temperature, or areas with little temperature seasonality, low mean annual precipitation, low soil clay fractions or high seasonality in precipitation (Du et al., 2020). These factors help determine the ability of soils to contribute to agricultural productivity and the degree to which individual crop species may succeed.

The internationalization of markets for agricultural output has led to the transport of soil nutrients from the site of origin to regions that are different to a large extent in geology, climate, biota and soil quality. Trade in agricultural products doubled between 1995 and 2018 (FAO, 2020). It has been estimated that the movement of associated soil nutrients is costly and increasing. Grote et al. (2005) estimated that the movement of N, P and potassium (K) doubled to 8.8 Tg between 1997 and 2020. It was later proved that the movement of the four dominant staple crops (maize, wheat, rice and soy) increased 2.3 times between 1997 and 2019, indicating that embedded nutrients likely increased at a similar rate (FAOSTAT, 2021). Some of the nutrients from soils engrossed in agricultural products and transported via national and international sales will be immobilized in local biomass for years to decades. Nevertheless, part of the nutrients transported will end up in food waste within weeks to months after arrival. About 194 to 389 kg y^{-1} of food is wasted per person globally (Corrado and Sala, 2018), amounting to an estimated one-third of food production (Gustavsson et al., 2011). Nutrients imported by way of food and feed that ultimately finish up as food waste are rarely applied to depleted soils. Rather, these nutrients tend to concentrate in urban locations where they contribute to greenhouse gas emissions, eutrophication and other forms of contamination (Grote et al., 2005; Crippa et al., 2021).

2.1. FOOD AND FEED FOR A GROWING GLOBAL POPULATION

The World's population is expected to increase by approximately 2 billion people by 2050 and reach approximately 11 billion in 2100 (UN, 2019). Providing sufficient food for this growing worldwide demographic is a huge hurdle of the current century. Augmented crop yield will be required to provide enough food and feed over both the immediate and more distant future (Mueller, 2012). This necessitates higher agricultural production, placed against a context of widespread and increasing land over utilization (Bai et al., 2008; Montanarella et al., 2015) and increasing challenges from climate change (van Meijl et al., 2017). It has been predicted that agricultural output will require a three-fold increase by 2100 to match worldwide demand under a regular scenario (Beltran-Peña et al., 2020), with higher agricultural output required during the next 30 years than has been managed over the previous 400 years Keating et al., 2014). Impending shortcomings in food-related energy and nutrients are worsened by increasing inefficiencies in the utilization of soil resources. For example, meat and dairy intake, escalating in richer countries, generally has lesser efficiencies of conversion of energy from plants into edible protein than consumption of non-meat food sources (Smil, 2002; Tilman et al., 2011). Cattle remain a significant source of food but are a toll on global food energy if fed with potential food crops. Nevertheless, cattle are a net gain to the food system when fed with grass (aka pasture raised) or by-products that would not otherwise be put to use for human consumption (Smith, 2013). For example, shifting 16 major crops to 100% food rather than as feed for animals would augment available food energy by 49% worldwide (Foley et al., 2011). A change in energy intake from animal to crops may go against some cultural practices and present other hurdles to adoption, yet at an international scale may have an impactful outcome on meeting future food requirements.

Although food production will require an increase in all food-producing regions to match increased demand, the expected level of food self-sufficiency differs significantly by region, with the minimal estimates in Africa and Southeast Asia. Lacking food in these regions is due in part to inherent soil and climatic conditions, in addition to patterns in population development that surpass projected growth in crop yields using current approaches, in tandem with socio-economic and political forces creating inequality in food and agricultural resource distribution (Beltran-Peña et al., 2020; D'Odorico et al., 2019).

2.2. THE RELATIONSHIP OF SOILS TO FOOD AND FEED

2.2.1. The Main Soil Types in Worldwide Agriculture

From an ecological point of view, the potential yield of agricultural land is strongly dictated by the combination of naturally occurring soil chemistry, physical characteristics and climate, all of which are covered by soil classification. Particular soil forms tend to dominate worldwide food and feed output when controlling for climate, owing to a combination of physical and chemical characteristics that more forcefully promote plant development. Mollisols are one of the most intensively farmed soils worldwide, especially in the Americas, Europe and Asia (Hatfield, 2017). Mollisols are likely to be known for the accumulation of surface soil organic material, which adds to the nutrient pool and to nutrient and water upkeep. Mollisols also tend to be found in stable landscapes missing steep topographic variation and thus are less impacted by erosion than some other common soil types (Hatfield, 2017; Eswaran, 2012). Likewise, Alfisols in Africa and Europe, and some Inceptisols in Europe and South Asia are heavily farmed due to inherently favorable nutrient levels derived from either high soil organic material amounts or input of abundant weathering products (Hatfield, 2017; Eswaran, 2012). Mollisols and Alfisols combined make up about 17% of the global land surface, while Inceptisols cover about 15% of the global land surface.

Entisols are usually known for shallow profiles, steep slopes and other easily erodible surfaces, and are likely to support reduced agricultural productivity in the absence of inputs, with the possible exception of Entisols developed from recent alluvial deposits (Grossman, 1983). Entisols cover approximately 18%

of the global land surface and are dominant in Africa and parts of Australia (Eswaran et al., 2012). Aridisols (12% of the worldwide land space), or dry land soils, happen under generally non-arable conditions. These soils are likely to be home to grazing activities where agricultural plants cannot grow. Irrigation can improve plant development in certain Aridisols, but high salt content stops plant development in others (Nettleton and Peterson, 1983). Aridisols in regions of North America, South Africa, Kazakhstan, Australia and Argentina are dominated by cattle grazing activities (Eswaran et al., 2012; Asner et al., 2004). Andisols, derived from recent volcanic activity, cover under 1% of the Earth's land area and can support particularly fertile soils. Deforestation for agricultural production is a growing concern on Andisols in tropical areas (Samll and Naumann, 2011).

2.2.1.1. Agricultural Soil Nutrients

Soils vary strongly in their inherent nutrient pools. A lot of the cropland with the highest native soil nitrogen contents globally, which also have some of the greatest crop yields, mainly produce cattle feed or biofuel (Cassidy et al., 2013; ISRIC, 2020). In 2019, under 10% of the energy production in agriculture in the USA was utilized for human intake, with most of the energy loss due to inefficiency in nutrient conversion of corn and soy to food calories or animal meat via use as biofuel or feed for cattle (USDA, 2020). Regardless of lower overall soil nutrient availability and productivity, agriculture in eastern and central sub-Saharan Africa and India have among the greatest overall rate of conversion of crop-embedded nutrients and energy sources to human food worldwide. Crops contributing to high food calories include corn, wheat and cassava in southern Africa; sorghum, sugar crops, teff and roots and tubers grown in east Africa; and sugar crops, rice and legumes produced in India (FAOSTAT 2021; Cassidy et al., 2013; FAO, 2020). The strong impact of economics on agricultural areas is palpable in Indonesia, which has some of the greatest crop productivity globally, yet under half of the energy produced is converted to food due to the dominance of palm oil cultivation (Condro et al., 2020).

Handling of soil nutrients is a determinant factor for the productivity of crop lands. Abuse of soil nutrients by way of harvest lacking enough replacement via fertilizer inputs, lack of cover cropping, compaction from heavy machinery and salinization linked to irrigation are common causes of soil degradation and reduced agricultural productivity (Tan et al., 2005; Bai et al., 2008). Nutrient depletion alone impacts over 130 million hectares of food growing lands, approximately 8% of the global agricultural area and over 25% of farmland in Latin America (Bai et al., 2008). About 49% of food produced globally is carried out outside existing planetary boundaries (Rockström et al., 2009), with 25% of this originating from the over-exploitation of soil nitrogen resources (Gerten et al., 2020). Technological developments in fertilizer production, pesticides and genetically modified cultivar advances have grown the productivity of croplands for land workers with economic means, although at the expense of environmental impacts (Carvalho, 2006). When looking at the yields of the top internationally traded crops, Mueller et al. (2012) found that nutrient and irrigation practices were the main differential factors in potential crop gaps, while coarse-resolution soil organic carbon and qualitative soil texture as factors did not explain variation tendencies (Mueller et al., 2012). Fertilizer utilization has increased by approximately an order of magnitude since 1950, with a current international use of 109 Tg N $y-1$ (Steffen et al., 2015; Kopittke et al., 2019). After 1970, the majority of the growth in fertilizer utilization took place in developing countries (e.g., BRICS countries; Steffen et al., 2015), which have strengthened their participation in global food sales. The significant growth in nitrogen introduced globally, and the separation of nitrogen from ecological processes (Vitousek et al., 1997), have created chaos on water, soil and atmospheric chemistry (Lassaleta et al., 2014; Nesme et al., 2018; Oita et al., 2016; MacDonald et al., 2015).

As with nitrogen, soil phosphorus is key for plant growth and levels of agricultural phosphorus fertilization have gone up significantly at a worldwide scale (Nesme et al., 2018). Soil phosphorus bioavailability has gotten limited attention, although it may reduce crop productivity by 20–50% in some areas (Kvakic, 2018). Phosphorus often has a capped level of bioavailability in soils, especially in the acidic soils of the tropics; crop species with development strategies designed to pull out soil phosphorus can be more successful in these locations (Batjes, 2012). With the augmented utilization of irrigation and inorganic fertilizers to improve soil nutrient and water levels, agricultural output, and thus the food system, have become ever more detached from local soil biogeochemical systems.

2.2.1.2. The Role of Soil Characteristics in the Human Nutrition of Food and Feed

Historically, it has been accepted that soils play a key function in human nutrition. References to the impact of soil quality for human health are present in religious texts dating back to 1400 BCE (Brevik and Sauer, 2015). Soils vary vastly in their natural nutrient availability and therefore their ability to provide nutrition to food. This is evident when examining local and international patterns in the prevalence of undernourishment. In 2019, an estimated 9% of the global population, or 690 million people, were affected by undernourishment (FAO et al., 2020), and food insecurity and associated patterns in poor nutrition have been going up at an international scale since 2014.

Although there are numerous reasons for undernourishment and food insecurity (e.g., extreme climate events, socio-political and economic conundrum), part of the world's nutritional insufficiency springs from soil nutrient limitation. The augment in food production over the past 50 years has made way for a reduction in some nutritional deficiencies, yet Ca, Fe, vitamin A and Zn are still broadly lacking (Beal, 2017). Micronutrients are of special concern. The worldwide growth of food production has driven up rates of soil micronutrient mining that often lead to eventual reduction in crop output (Lal, 2015). This pattern is mostly seen in developing countries where fertilizers are regularly economically unavailable (Sanchez and Swaminathan, 2005). As is the case of human zinc deficiency, it correlates with soil zinc deficiency in sub-Saharan Africa, some areas of South America and South and Southeast Asia (FAO et al., 2014; Cakmak et al., 2017). Micronutrient deficiency is a leading contributor to worldwide disease burden, affecting an estimated 50% of the global population (WHO, 2002). Low levels of soil Zn, Cu and Mn, specifically, have been connected to increased child mortality (Berkhout et al., 2019). Micronutrient deficiencies are more often found in locations where crops have low nutritional content and are a dietary staple (Brevok et al., 2020), and are more likely to happen in areas with dependence on native pollinators (i.e., vitamin A in South Asia and Fe in sub-Saharan Africa) and with the disruption of plant-pollinator processes (Chaplin-Kramer et al., 2014).

Climate change additionally limits the ability of soil to offer essential nutrients for food and feed. As an example, crops grown under elevated CO_2 had 3–17% less Zn, Fe and protein amounts in their composition. This is especially present in C3 staple crops (rice and wheat) and corn (being the only C4 crop affected) (Myers et al., 2014). Protein in staple foods is expected to be reduced by 7.6–14.1% by 2050 as an outcome from elevated CO2 (Medek, 2017), with the greatest effect in sub-Saharan Africa and South Asia, including India (Medek, 2017). Nutritional vulnerability linked to elevated CO2 will have a disproportionate effect on low-income populations, who depend on vegetable food sources for the majority of their nutritional intake (Smith and Meyers, 2017).

Soil degradation is present in the majority of the world's crop land, in both intensive commercial as well as subsistence agriculture (Alloway, 2008), reducing the *per capita* nutritional value of food worldwide (Beal et al., 2017; Schmidhuber et al., 2018). Reversing the current tendencies in declining soil fertility will need the setting up of soil nutrient stocks via strategies such as soil liming, soil organic matter addition and micronutrient fortification (Lal, 2015; Wood and Baudron, 2018; Dimkpa and Bindraban, 2016). A delineation of soil active zones at local and international levels in which soil physico-chemical composition is dictated spatially would help the development of specific macro- and micronutrient strategies for sustainable food production and soil betterment (Behera et al., 2020).

CHAPTER 3

WHAT IS A HEALTHY SOIL?

Contents

Soil health is the concept used for the ability of soil to work as a living system. Healthy soils keep a varied community of soil organisms that aid with the management of plant disease, insect and weed pests, create beneficial symbiotic links with plant roots, recycle essential plant nutrients, improve soil structure with positive effects for soil water and nutrient holding capabilities, and ultimately improve yield. A healthy soil additionally helps with the mitigation of climate change by keeping or increasing its carbon content.

Food availability depends on soils. Nutritious and high-quality food and animal fodder are only produced if the soils are in healthy condition. A healthy living soil is thus a key factor to achieve food security (Figure 3.1).

Figure 3.1. *Ecosystem functions supported by healthy soils.*

Over the past 50 years, development in agricultural technology has made way for a significant step in food production and boost in the global food security status. Nevertheless, in numerous nations, this strong crop management has stripped the soil, jeopardizing the maintenance of production in these regions in the future. With a worldwide population that is expected to surpass 9 billion by 2050, in addition to competition for land and water resources and the impact of climate change, current and future food security depends on the possibility to increase crop output and food quality utilizing the soils that are in current production.

Various and diverse farming methodologies promote the sustainable usage of soils with the objective of optimizing management, for example: agroecology, conservation agriculture, organic farming, zero tillage farming and agroforestry.

Agroecology makes use of ecological theory to investigate and apply agricultural systems so as to make them both increase in productivity and improve at maintaining natural resources. This entire systems approach to agriculture and food systems development is founded on a ranging variety of technologies, practices and innovations, including regional and traditional knowledge in addition to modern science. By comprehending and working with the interactions between plants, animals, humans and the environment within agricultural systems, agroecology covers multiple dimensions of the food system, including ecological, economic and social.

Agroecology is a well-rounded and integrated methodology that applies a combination of ecological and social theories and principles to the creation and handling of sustainable agriculture and food systems. It aims to optimize the link between plants, animals, humans and the environment while also solving for the need for socially equitable food systems within which humans can exercise selection over what they eat and how and where it is made. Agroecology is at the same time a science, a set of practices and a social movement which has developed as a concept over recent decades to expand in coverage from a focus on fields and farms to include the entirety of agriculture and food systems. Agroecology now represents a transdisciplinary field that encompasses the ecological, socio-cultural, technological, economic and political dimensions of food systems, from production to intake.

It is now impossible to consider food, livelihoods, health and the management of natural resources apart from each other. Adopting systems–thinking through all-encompassing approaches is required to face these multifaceted and codependent issues. The key link between individuals and Earth, with sustainable agriculture and food systems, is at the center of the 2030 Agenda for Sustainable Development, which emphasizes the pressing requirement to mobilize and pursue policies aimed at transformational modifications. Ending poverty and achieving zero hunger, while guaranteeing inclusive development and sustainably managing the Earth's natural resources, all in the presence of climate change and biodiversity reduction, will only be feasible through holistic and integrated approaches that consider human rights.

Agroecology stands for an overarching and comprehensive systems framework to be used as reference for public policies aiming for sustainable agriculture and food systems. It strengthens public efficiency by fostering integrated and inter-ministerial policy design and implementation, connecting agricultural and food sectors that are often separated. It actively brings together different stakeholders through inter-disciplinary steps which are setup to achieve a responsible and transparent governance of resources. In this way, agroecological transitions may support the simultaneous achievement of multiple sustainability targets – economic, environmental, social, nutritional, health and cultural – holistically and in a connected form at different stages and scales while being adapted for different environmental and cultural settings.

Agroecology is designed on bottom-up and territorial procedures, helping to deliver contextualized answers to local issues with people at the core.

3.1. THE GLOBAL MOVEMENT OF SOIL NUTRIENTS VIA FOOD AND FEED EXPORT

The contribution of soil to food and feed covers many spatial scales from less than 1 km for food and feed produced regionally to over 10,000 km for transcontinental global trade. International trade in food and feed have been drastically modified during the second half of the twentieth century and into the beginning of the twenty-first century (Steffen et al., 2015), with the outcome of large-scale movement of soil nutrients accumulated in agricultural production. Between 1961 and 2009, international crop trade augmented the movement of plants and animal proteins by a factor of 7.5 and 10, each (Lassaletta et al., 2016). Livestock accounted for an estimated 75% of global agricultural yield in 2009, with China, North America, Europe and Oceania as the main consumers; animal feed made up the largest contribution to exported nitrogen (Lassaletta et al., 2016). Nitrogen found in animal production went up by 60% from 1986 to 2009, while the jump in per capita nitrogen intake and consumption of animal protein both went up by only 12% ((Lassaletta et al., 2014). Between 1960 and 2011, embedded phosphorus in internationally traded food and feed went up 7.5 times, which made way for an expansion of agricultural phosphorus flows for human intake (28%), animal feed (44%), and other agricultural plants (11%) (Nesme et al., 2018). The majority of the soil phosphorus carried via food and feed trade started in the Americas. Net phosphorous imports took place in regions with significant populations (Asia), large livestock rearing (Europe) or with strong food security challenges (Mediterranean basin) (Nesme et al., 2018).

The four main staple crops, corn, soy, rice and wheat, are widely traded and exported internationally, transporting the calories and nutrients contained within the harvested plants. From 2014 to 2018, North and South America were the largest exporters of soy and maize crops. Soy exports caused a global transport of 3.2 ± 0.1 MMT N y^{-1} and 0.4 ± 0.02 MMT P y−1 from North America. South America exported 4.3 ± 0.3 MMT N y−1 and 0.6 ± 0.05 MMT P y−1 in soy. For context, this was equal to 58% and 23% of the total agricultural nitrogen fertilizer use on all crops in South and North America, respectively, over the same interval. The high nitrogen content in soy derives minimally in part from biological nitrogen-fixation, but nitrogen export from these high exporting regions shows the decrease of potential soil nutrients. Phosphorus export in soy from these locations was elevated as well, equal to an estimated 9% of the total phosphorus fertilizer used in these areas. During the 5-yr period, 14 ±

1% and 58 ± 4% of soy exported from North and South America, respectively, was utilized for animal feed. South America strongly dominated the soy export market for animal feed, moving 3.2 ± 0.1 MMT of soil nitrogen y−1 and 0.43 ± 0.01 MMT of soil phosphorousu y−1 globally (Liu et al., 2017).

Although soy cultivation started in China, Chinese farmers targeted corn and rice over the twentieth century owing to higher total financial returns per hectare on crops apart from soy (Lassaletta et al., 2016). The combination of modifications in Chinese domestic and foreign policies in the 1990s and increasing livelihood levels in China prompted rapid development in soy feed and oil intake from basically no soy imports in 1990 to making up approximately 65% of international soy imports in 2018 (Gale et al., 2019). Brazil's increasing land utilization, ability to double-crop soy with corn in its tropical climate, and USA–China trade-tariff policies pushed Brazil's soy production from 10 million ha in 1980 to 35 million ha in 2017. This increase has primarily taken place on nutrient-poor Oxisols of the Cerrado region through economic and subsidy aid, implementation of fertilizers and liming, and implementation of suitable cultivars. In tandem, the decrease of soy production in China and its substitution with N-intensive crops have conducted to a change in the nitrogen balance from negative (soy) to increasingly positive, with about half of the contribution (49%) linked to cropland conversion and higher mineral nitrogen fertilizer application (51%) (Sun et al., 2018).

Annual nitrogen and phosphorous transport of corn was an order of magnitude lower than soy from these areas, with average corn exports displacing 0.82 ± 0.07 MMT N y−1 and 0.17 ± 0.01 MMT P y−1 from North American soils between 2014 and 2018, and 0.65 ± 0.04 MMT y−1 and 0.13 ± 0.01 MMT P y−1 from South American soil. Corn export was equal to an estimated 6% and 9% of the regional-scale total nitrogen fertilizer utilization for North and South America, respectively. Asia was the biggest importer of these embedded nutrients. Animal feed only makes up 3.5 ± 0.2% and less than 0.1% of corn exports from North and South America, respectively, mostly as corn meal.

Asia and Europe were the biggest importers of soy and corn meal for animal feed. Adding up the export of soy and cornmeal with the contained nutrients in the export of livestock meat was responsible for a total global movement of approximately 3.9 ± 0.1 MMT nitrogen y^{-1} and 0.53 ± 0.01 MMT phosphorus y^{-1}, a value equal to the total nitrogen fertilizer utilized in Africa and 43% of the total phosphorus fertilizer utilized in Oceania. Global export of livestock meat embedded only 0.33% of the nitrogen and 0.15% of phosphorus y^{-1} exported as soy and corn at an international scale. Oceania had the largest soil nutrient export embedded in cattle meat at 4.1 ± 0.2 KT nitrogen y^{-1} and 0.2 ± 0.01 KT phosphorus y^{-1}.

3.2. The 10 Elements of Agroecology

The 10 Elements of Agroecology resulted from a multi-stakeholder process intended to generate a system re-design framework to be optimized and adapted to local contexts. The 10 Elements of Agroecology framework was developed between 2015 and 2019. Prominent themes from presentations delivered during the First International Symposium on Agroecology for Food Security and Nutrition provided an initial coherent structure: recycling, efficiency, diversity, resilience and synergies as central ecological features of agroecology. Nevertheless, calls in regional meetings for reinforcing social and political aspects of agroecology were also strong. Thus, these aspects that emerged from regional consultations were clustered under an additional five elements: co-creation of knowledge; human and social values; culture and food traditions; responsible governance; and circular and solidarity economy. Following the refinement of element names, content, and the development of a consistent storyline highlighting the interconnected nature of agroecology and its 10 Elements, the framework was finalized after several rounds of review by international and FAO experts.

The 10 Elements of Agroecology framework was launched at the Second FAO International Symposium on Agroecology held in April 2018 and continues to evolve. In October 2018, the 10 Elements of Agroecology were supported by the FAO Committee on Agriculture (COAG) at its 26th Session as a guide to one of the ways to promote sustainable agriculture and food systems. Following the review, revision, and clearance process through FAO's governing bodies, the 10 Elements of Agroecology were approved by the 197 Members of FAO to guide FAO's vision on agroecology by the 163 session of the Council on 2–6 December 2019.

- **Diversity:** diversification is key to agroecological transitions to ensure food security and nutrition while conserving, protecting and enhancing natural resources.
- **Co-creation and Sharing of Knowledge:** agricultural innovations respond better to local challenges when they are co-created through participatory processes.
- **Synergies:** building synergies enhances key functions across food systems, supporting production and multiple ecosystem services.
- **Efficiency:** innovative agroecological practices produce more using less external resources.
- **Recycling**: more recycling means agricultural production with lower economic and environmental costs.
- **Resilience:** enhanced resilience of people, communities and ecosystems is key to sustainable food and agricultural systems.

- **Human and Social Values:** protecting and improving rural livelihoods, equity and social well-being is essential for sustainable food and agricultural systems.
- **Culture and Food Traditions**: by supporting healthy, diversified and culturally appropriate diets, agroecology contributes to food security and nutrition while maintaining the health of ecosystems.
- **Responsible Governance:** sustainable food and agriculture requires responsible and effective governance mechanisms at different scales – from local to national to global.
- **Circular and Solidarity Economy:** circular and solidarity economies that reconnect producers and consumers provide innovative solutions for living within our planetary boundaries while ensuring the social foundation for inclusive and sustainable development.

Organic farming refers to agricultural production leaving out the utilization of synthetic chemicals or genetically modified organisms, growth modifiers, and additives to livestock fee. It also highlights a holistic farm management process, where rotations and animals have a key role on the system. Soil fertility is the key of organic farming. Since organic farmers skip synthetic nutrients to fix degraded soil, they are required to concentrate on setting up and maintaining soil fertility mostly through their basic farming practices.

Organic farming is an agricultural practice with the objective to produce food employing natural substances and processes. Thus, organic farming is likely to have a limited environmental load as it aims to achieve the following:

- responsible use of energy and natural resources;
- maintenance of biodiversity;
- preservation of regional ecological balances;
- enhancement of soil fertility; and
- maintenance of water quality.

In addition, organic farming practices encourage an elevated standard of animal welfare and ask farmers to meet the specific behavioral requirements of animals.

The regulations of the European Union on organic farming are created to give a clear structure for the development of organic goods throughout the entire EU. This is to meet the demand of consumers for reliable organic products while providing a suitable marketplace for growers, distributors and marketers.

In order for growers to obtain benefits from organic farming practices, consumers must trust that the guidelines on organic production are being observed. Thus, the EU has set up the following strict system of management and enforcement to ensure that organic rules and regulations are applied correctly. As organic farming is a component of a larger supply chain which covers food processing, distribution and retail sectors, these are additionally subject to verification.

- Each EU member is required to set up 'control bodies or authorities' to check procedures in the organic food chain. Producers, distributors and marketers of organic products have to register with their regional control body prior to being allowed to sell their food as organic.
- Following a control and check, they will be given a certificate to corroborate that their products match organic standards.
- All operators are evaluated at least once a year to ensure that they are continuing to follow the organic production regulations.
- Imported organic food is additionally subject to control management to ensure that they have also been grown and shipped according to organic guidelines.

3.2.1. The Organic Logo

The organic logo provides a coherent visual identification to EU produced organic goods sold in the EU. This creates an easier method for EU-based users to identify organic products and aids producers to sell them across all EU member states. The organic logo may only be used on goods that have received certification as organic by an authorized control agency or body. This translates into having met strict conditions on how they are grown, transported and stored (Figure 3.2).

Figure 3.2. *The organic logo.*

The logo is mandatory for the majority of organic products and is required to be displayed in accordance to a specific set of rules. This is to avoid consumer confusion, help with the maintenance of trust in organic goods and help the authorities in their inspection routines.

The logo must be used by:

- all pre-packaged EU foods, produced and sold as organic within the EU.

In addition, the logo can optionally be used by:

- imported products where the product meets the EU rules on the import of organic goods;
- non-pre-packaged organic goods;
- EU organic products to be sold on third country markets; and
- as part of information campaigns aimed to educate the public about the organics program (as long as it is not misleading or utilized to imply that a non-organic good meets the requirements of an organic good).

The logo may not be used for:

- products containing under 95% organic ingredients;
- mass catering operations including restaurants or hospitals;
- products outside the scope of organic rules including cosmetics or products from hunting and fishing; and
- products in 'conversion' (where organic methods have only just been introduced and there may still be non-organic substances in the soil or animal chain).

3.3.2. Organic Farming After 2022

Organic farming is a rapidly growing practice in European agriculture, as a direct result of greater consumer interest in organic products. As an answer to the challenges presented by this quick expansion, and so as to give an effective legal framework for the industry, the European Union passed new laws addressing the organic sector which went into effect from 1st of January 2022.

Examples of the alterations that were made under the new organic laws include:

- a fortification of the control system, helping to increase further consumer confidence in the EU organics system;
- new rules for producers which will facilitate smaller farmers to change to organic production;
- new rules on imported organics to guarantee that all organic products marketed in the EU are of the same standard;
- a larger range of products that can be sold as organic.

The new organic legislation is backed up by the action plan for organic production in the European Union, which was launched by the European Commission in March 2021.

New organic legislation is applicable from the 1st of January 2022, after the postponement of its implementation for a year. The changes to legislation reflect the changing nature of this quickly growing sector. The new regulation is created to guarantee fair competition for food producers whilst halting fraud and keeping consumer trust through the following:

- production rules are made easier through the phasing out of a number of exceptions and opt outs;
- the control system is made stronger thanks to tighter precautionary measures and robust verifications along the entire supply chain;
- producers in third countries will have to meet the same set of rules as those producing in the EU;
- organic rules cover a wider list of products (e.g., salts, cork, beeswax, wool, etc.) and have additional production rules (e.g., deer, rabbits and poultry);
- certification will be facilitated for small farmers thanks to a new system of group certification;
- there will be a more uniform approach to lowering the risk of accidental contamination from pesticides.

Box 1: Case Study: Organic Farming in Peru

The major blockage to augmenting agriculture production in Palau is missing appropriate sustainable technologies to increase the quality of soil for crop production. Palau's soils are highly weathered, acidic and clayey and there is a capped amount of land area that is fitting for agricultural production without substantial inputs and soil modifications.

The Food and Agriculture Organization is assisting the Bureau of Agriculture (BOA) in strengthening the Palau Organic Growers Association (POGA) to augment domestic agricultural output through taking up new sustainable technologies and practices for soil improvements.

FAO is making training accessible for POGA farmers and BOA extension staff on fundamental aspects of soil biology, soil testing, fertility improvement strategies, demonstration trials and other sustainable organic production technologies and crop management. The FAO is also backing up farm demonstrations with selected farmers to adapt technologies to regional conditions and major crops.

Source: FAO (2015).

CHAPTER 4

CONSERVATION AGRICULTURE

Contents

Conservation Agriculture is a set of farming practices that promote minimum soil disturbance (e.g., no tillage), preservation of a permanent soil cover, and diversification of plant types. It strengthens biodiversity and natural biological processes above and below the ground surface, which add to greater water and nutrient utilization efficiency and to improved and ongoing food production. Conservation agriculture methodologies have greatly improved soil conditions, lowered land degradation and increased yields in various parts of the globe by applying three principles:

- minimal soil disturbance
- permanent soil cover
- crop rotations

To be sustainable in the long term, the reduction of organic matter in any agricultural ecosystem must never be above the rate of soil formation. In most agro-ecosystems, that is impossible if the soil is mechanically altered. Thus, one of the tenets of conservation agriculture is controlling the use of mechanical soil disturbance, or tilling, in the farming practices.

Conservation agriculture, according to the definition of the United Nations' Food and Agriculture Organization, is "a farming system that promotes maintenance of a permanent soil cover, minimum soil disturbance, and diversification of plant species. It enhances biodiversity and natural biological processes above and below the ground surface, which contribute to increased water and nutrient use efficiency and to improved and sustained crop production." The special report "Climate Change and Land" (2019) includes conservation agriculture as part of the incremental adaptation options to address climate risks. The three main principles of conservation agriculture (minimum soil disturbance, crop diversification, and permanent soil cover) aid in the protection of the environment and to lower both the impacts of climate change on agricultural systems (adaptation) and the contribution of the agricultural practices to greenhouse gasses emissions (mitigation) through sustainable land management. These principles, described in more detail below, aid in the protection of the soil from erosion and degradation, increasing soil quality and biodiversity, preserve the natural resources and augment their use efficiency, while optimizing food yields.

More in details, "minimum soil disturbance" is known for reduced tillage practices (including ploughing, harrowing, and all the tillage operations ordinarily applied to prepare the soil for the germination of seeds, seedling establishment and crop growth and production) through direct seeding and/or direct fertilizer placement. It helps to better soil properties, maintain and increase soil organic matter, and thus reduce soil erosion. In addition, no tillage and minimum tillage lower energy consumption by agriculture machinery, increase soil drainage, strengthen food supplies for insects, birds, and small mammals because of higher availability of crop residues and weed seeds in the soil. In that way a number of ecosystem services are obtained by the minimum soil disturbance, including: water regulation, carbon storage, soil stability, protection of surface soils from erosion, strengthened water infiltration, greater soil fertility through enhanced nitrogen stocks (in the long term), improved soil, water, and air quality, less soil erosion and fuel use. All these components are of the highest importance so as to reduce the vulnerability of the agricultural systems and strengthen their adaptation capacity to climate change, contributing also to the mitigation targets.

"Crop diversification" is the method of cultivating more than a single species in a given agricultural section, by making use of crop rotation and/or association. The diversification in cultivated species augments the adaptation capacity of agricultural systems to climate change by enhancing soil fertility and structure, soil water holding capacity and water and nutrients distribution through the soil profile, helping to avoid pests and diseases, and augmenting yield stability. Thus, the diversified cropping systems are more stable and resilient than monoculture systems. Crop diversification carries a range of ecosystem services, contributes to better crop productivity and resilience of farming systems and lowers greenhouse gasses emissions from agricultural practices.

"Permanent soil organic cover" with crop residues and/or cover crops (including legumes, cereals, or other crops planted between the main crops, mainly for the benefit of the soil rather than the crop yield) enables climate change adaptation by reducing soil erosion and degradation which can be made worse by the impact of extreme weather events (including extreme precipitations, droughts and periods of soil saturation, extreme heat, strong wind events) and improving the stability of the conservation agriculture system. Cover crops enhance soil properties (fertility and quality), aid in the management of soil erosion, preserve soil moisture, avoid compaction of the soil, contain pests and diseases, and enhance biodiversity in the agro-ecosystem.

4.1. STAKEHOLDER PARTICIPATION

A successful setup of conservation agriculture needs stakeholder participation from both public and private sectors and a meaningful collaboration between the various sections involved: farmers, farm advisory services (that provide farmers knowledge and skills to improve the applied agronomic techniques, crop productivity and farm income), researchers, policy makers, among others. Effective stakeholder-based participatory approaches are required to guarantee the dissemination and application of conservation agriculture practices and to polish the measures in accordance to the specific characteristics of the agricultural systems considered, to gain the greatest possible effectiveness. Farmers and other stakeholders must be engaged in projects dealing with conservation agriculture mechanisms, to obtain more awareness of the close connection among agricultural practices, environmental impacts and socioeconomic effects, covering the potential for adaptation to and mitigation of climate change.

Additionally, farmers should be helped during the starting period of change from traditional to conservation agriculture, to gain all the needed information and to obtain experience with the new practices and be conscious of the work and time required for the transition to the new cultivation system. In this setup, the role of farm advisory services is key, as well as the improvement of capacity building and education. The presentation of the effects of conservation agriculture techniques applied on real case studies could aid the setup of the measures and give indications to new farmers on which paramount practices create success and what mistakes to avoid.

4.1.1. Success and Limiting Components

The success components for the setup of conservation agriculture measures include: good stakeholder engagement, policies and government actions to promote and create favorable conditions for the setup of conservation agriculture (including free access to information), appropriate farm advisory services, public and private partnerships, and rewards for environmental services.

Some aspects can perform as limiting components for small farm dimensions, as for example for the implementation of practices that need investments in machinery (as for sod seeding in non-tillage farming systems). In these cases, organizations of farmers or collaboration with third parties are utilized to overcome this aspect. Other limiting factors are the inadequate dissemination of knowledge and good practices, the limited collaboration between researchers and farm advisory services and the missing support for farmers.

At times, a perception from the farmers remains that tillage is required for soil improving, facilitating crop management and obtaining higher yields. Additionally, farmers are generally pleased with the actual practices and do not see an economic incentive to change, as clean and well-tilled fields are regularly connected with good farming practice. In this regard, the farm advisory services play a fundamental role in encouraging the confidence of farmers recently acquainted with conservation agriculture reassuring them that the technology is working. This includes showing the technology in other farmers' fields, demonstrating the economic advantages with facts and numbers and training people in the region to help others.

4.1.2. Costs and Benefits

The cost for setting up conservation agriculture measures is likely to oscillate across farms (depending on size and production system), geographical locations, and countries. Nevertheless, it is reported by the Food and Agriculture Organization that by not tilling the soil, farmers can save between 30% and 40% of time, labor and, in mechanized agriculture, fossil fuels as compared to traditional agriculture practices, lowering the connected costs. Overall, conservation agriculture permits a lowering of the production costs and a lowering of time and labor (including for example land preparation and planting), and in mechanized systems it lowers the costs of investment and upkeep of machinery in the long term. Additionally, it can achieve yields comparable with modern intensive agricultural methods but in a sustainable manner, allowing plants to better adapt to changed climatic conditions in regards to the ordinary farming management, particularly by reducing year-to-year yield variability. Nevertheless, the positive effects on food yields depend on the intensity and severity of climate change effects.

The economic, agronomic and environmental gains provided by conservation agriculture are apparent at global, regional, local and production level. These benefits are also relevant in regards to the adaptation to climate change, as through conservation agriculture crop yields are maintained or even improved, as well as in terms of mitigation, by achieving greater carbon sequestration and reducing greenhouse gasses emissions.

4.1.3. Legal Aspects

Conservation agriculture methodologies should be supported by clear policies and procedures. The Common Agricultural Policy (CAP) of the European Union and the National and Regional Rural Programs are among the main policy driving forces for the implementation of conservation agriculture in the European Union Member States.

The CAP promotes the application of these practices through the "green direct payment" (or "greening") (first pillar of the CAP) to aid farmers who adopt or keep farming practices (e.g., crop diversification) that assist in meeting environmental and climate goals. Additionally, the second pillar of the CAP, the European Union's rural development policy, designed to support the rural areas, allows regional, national and local authorities to develop their individual Rural Development Programs and prompts, among others, measures for sustainable management of natural resources and climate action, including the conservation agricultural practices. The second pillar programs are co-financed by European Union funds and regional or national funds.

4.1.4. Implementation Time

A year can be sufficient to implement the measures of conservation agriculture. The time needed is highly dependent on knowledge dissemination, policies and government involvement, skills and financial support availability, in addition to stakeholders' involvement (Figure 4.1).

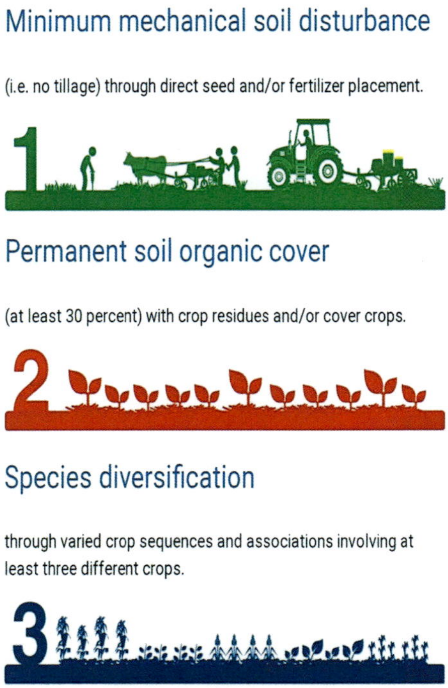

Figure 4.1. *Three principles of conservation agriculture.*

Source: FAO (2024).

To be widely taken up, all new technology requires benefits and advantages that attract a significant group of farmers who comprehend the differences between what they are doing and what they require. In the case of conservation agriculture these benefits may be categorized as:

4.1.4.1. Three Major Economic Benefits Resulting from Conservation Agriculture Adoption

- Time saving and therefore reduction in labor needs.
- Reduction of costs including fuel, machinery operating costs and maintenance, in addition to a reduction in labor cost.

- Higher efficiency in the sense of more output for a lower input.

The positive effects of conservation agriculture on the distribution of labor during the production cycle and, more impactfully, the reduction in labor requirement are the key reasons for farmers in Latin America to take up conservation agriculture, especially for farmers who rely fully on family labor.

4.1.4.2. Agronomic Benefits that Improve Soil Productivity

Taking up conservation agriculture leads to improvement of soil productivity including:

- Greater organic matter.
- In-soil water conservation.
- Improvement of soil structure, and from this rooting zone.

The constant inclusion of crop residues makes way to an increase in the organic matter content of the soil. To start with, this is reduced to the top layer of the soil, but with time this will expand to deeper soil layers. Organic matter plays a key role in the soil: fertilizer use efficiency, water holding capacity, soil aggregation, rooting environment and nutrient retention, all depend on organic matter.

4.1.4.3. Environmental Benefits That Protect the Soil and Make Agriculture More Sustainable

- Lower soil erosion, and thus lower road, dam and hydroelectric power plant maintenance costs.
- Improved water quality.
- Improved air quality.
- Greater biodiversity.
- Carbon sequestration.

Residues on the surface of the soil lower the splash-effect of the raindrops, and once the energy of the raindrops has dissipated the drops proceed to the soil without any negative impact. This results in greater infiltration and reduced runoff, leading to reduced erosion. The residues also create a physical barrier that lowers the speed of water and wind over the surface. A lower wind speed reduces evaporation of soil moisture.

One feature of conventional agriculture is its capability to change the landscape. The degradation of the vegetative cover impacts the plants, animals and microorganisms. Some few benefit from the change and turn into pests. Nevertheless, most organisms are negatively impacted and either they disappear completely or their amounts are drastically lowered. With the conservation of soil cover in conservation agriculture a habitat is developed for a number of species that feed on pests, which in turn attracts more insects, birds and other animals. The crop rotation and cover crops limits the loss of genetic biodiversity, which is favored with mono-cropping.

Systems based on high crop residue addition and no tillage, collect more carbon in the soil, compared to the loss into the atmosphere resulting from plough-based tillage. During the first years of setting up conservation agriculture practices, the organic matter content of the soil is augmented through the decomposition of roots and the contribution of vegetative residues on the surface. This organic material is decomposed slowly, and a lot of it is incorporated into the soil profile, thus the liberation of carbon to the atmosphere also happens slowly. In the total balance, carbon is sequestered in the soil, and turns the soil into a net sink of carbon. This could have impactful consequences in the fight to lower greenhouse gas emissions into the atmosphere and thus help to forestall the catastrophic impacts of global warming.

Box 1: Case Study: Conservation Agriculture in Lesotho

In response to the 2012 food insecurity crisis in Lesotho, the Food and Agriculture Organization and the Ministry of Agriculture and Food Security created a three-year cycle program to aid 18,500 households with agricultural inputs and education on agricultural technologies assisting communities to adapt to climate change, increasing their resilience to better withstand future shocks.

The program promotes conservation agriculture and greater home gardening and nutrition in all 10 districts of Lesotho. Soil erosion is a significant problem in Lesotho affecting both the quality and quantity of harvests. By taking up conservation agriculture, farmers not only guarantee better harvests but also contribute to the improvement of soil quality and its upkeep.

Subsistence farmers in Lesotho have been able to increase their agricultural yields and achieve greater food production by adopting conservation agriculture. The locally adapted practice, known as *likoti*, also helps to combat soil erosion and to enhance fertility. Since the mid-1970s, corn yields have decreased from an average 1400 kg/ha to a current 450–500 kg/ha in most of the districts. Regardless of its poor performance, agriculture remains a major source of livelihood for the vast majority of households dealing with subsistence farming.

Source: FAO (2015).

Zero tillage (ZT) is one of a set of methodologies used in conservation agriculture. Basically, it keeps a permanent or semi-permanent organic soil cover (either a growing crop or dead mulch) which protects the soil from sun, rain and wind and enables soil micro- organisms and fauna to take on the task of "tilling" and soil nutrient balancing – natural processes that are affected by mechanical tillage.

ZT is a key feature of conservation agriculture. It is extensively practiced in India, Bangladesh, and Nepal, where rice-wheat cultivation makes up the main cereal production system. The methodology has achieved much acceptance for its lower irrigation demand, prospects for improved soil fertility, and greater profit margins (Erenstein, 2009). Studies from India demonstrate improvement in soil organic content by 21% under the ZT option with the co-benefit of residue recycling. In the Basmati rice-wheat system, it augmented productivity by 36% and net returns by 43% compared to conventional practices (Jat et al., 2019). Nevertheless, maize systems produced a 5% bonus and potato crops provided a 70% bonus under ZT regimes as per studies from Bangladesh (Alam et al., 2015; Gathala et al., 2016). It is reported that ZT lowered irrigation needs by 27% and 22% compared to traditional methods for rice and wheat/cotton crops, respectively, while water productivity was improved 12-fold (Alam et al., 2015; Choudhary et al., 2016).

Against popular belief, no-till farming shows that agriculturalists can get perfectly fine sans plowing at the outset of field management. The idea of zero tillage predates conventional tillage by a long shot. Nevertheless, the reasons for not tilling today are different from those of previous millennia.

Historically, the lack or rudimentary nature of cultivation tools dictated the utilization of no-till practices. Currently, ecological concerns are mainly dictating the approach's implementation, thanks to the positive impacts of no-till farming on soil and the environment overall.

No-till farming, sometimes called zero tillage, zero-till, or direct drilling, is an agriculture approach that removes the need to disturb the soil through tillage. No-till cultivation needs specialized equipment (such as disc seeders or agriculture drills) to create furrows, plant seeds, and cover them—all at once, differently from double-passing the field after plowing. This way, the soil is subjected to limited disturbance because it is dug exactly where the seed is to be placed. Planters also facilitate the application of fertilizer directly into the furrow through specific tubes, resulting in limited fertilizer waste.

No-till agriculture needs greater weed control that includes herbicide applications. In conventional no-till farming, synthetic herbicide implementation is a frequent solution to eliminate invasive plants. Contrastingly, organic zero-till agriculture makes no use of any synthetic inputs. As an answer, no-till organic farming depends on integrated weed control strategies including crop rotation, cover crops mowed just before seed settles, grazing of livestock or poultry, and organic herbicides. By controlling weeds through no-till farming lacking herbicides and other sparing methods, agriculture producers add to environmental protection, nature regeneration, and human health.

The no-till methodology is key to regenerative farming, aiming to restore compromised soil to its starting condition. To increase soil organic matter and fertility, organic no-till farming promotes the following procedures (Figure 4.2):

- planting yearly or winter-hardy cover crops in the fall;
- leaving cover crops unharvested until the spring;
- killing the crops while at the same time transforming them into mulch with a roller crimper.

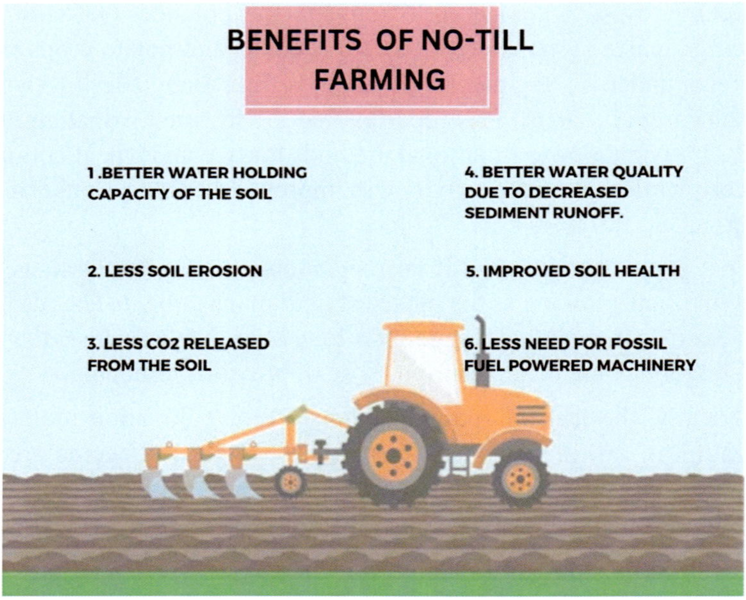

Figure 4.2. *No-tillage farming.*

To till or not to till – this is a frequently heard question that has been at the top of many agricultural debates. One of the strongest contributors to soil erosion is the simple (and greatly utilized) practice of tilling the land. The methodology has been around for thousands of years and was a turning point for modern farming systems. Yet, there is another way to farm. No-till farming is an approach that obtained great traction in the 70s and has slowly picked up strength. Despite its numerous advantages, it is not as two sided as it seems. For farmers looking to fix their land, choosing between tillage, no-till, herbicides and no-herbicides is a tightrope to walk.

4.2. UNDERSTANDING CURRENT PRACTICES

4.2.1. Tillage

Tillage is the process of preparing the soil for planting by digging it up, mixing it or turning it over. Tillage has often been conducted employing mechanical tools including a disc or plough that digs 25–30 cm into the soil. This process breaks up soil compaction, gets rid of weeds, and introduces air to the ground. The approach delivers short-term gains, but there are often long-term drawbacks.

The main consequence is tillage's effect on soil erosion. Tillage negatively impacts soil structure and aggregates, leaving it loose, exposed and vulnerable to erosion. This has significant side effects on the soil's water-holding power and ability to put up with extreme weather conditions.

Additionally, tillage also lets loose sequestered soil carbon back into the atmosphere, harms essential microorganisms, and is greatly resource-intensive.

4.2.2. Conventional No-Till

No-till farming is the practice of not tilling the land. The act uses tools that limit soil disturbance, including a direct seeder. There are many benefits, but the most notable is its ability to lower input costs.

Conventional tillage needs farmers to take several passes over the field, beginning with tilling the soil and then coming back to plant seeds. By practicing no-till, farmers can save heavily on labor, fuel and time. One study found that fuel expenses through no-till were lowered by 50–80%.

Currently, numerous conventional no-till systems depend on herbicides to get rid of crop residue and weeds. This creates another problem in the context of negative ecological impacts.

4.2.3. Herbicides and Glyphosate

There are numerous herbicides on the market. The most commonly used is glyphosate. The reason is that glyphosate is both cost-efficient and hugely effective in what it has been claimed to do.

However, its 'effectiveness' is also its greatest weakness. When glyphosate is sprayed, it enters the plant sap and spreads through the root system. As a non-selective herbicide, glyphosate kills any living plants it comes into contact with. As an outcome, it increased dependency on synthetic inputs and genetically modified plants, an increase in herbicide-resistant weeds, depleted biodiversity, detrimental effects on the health of people and polluted waterways. Glyphosate is usually neutralized and broken down by soil microbes after some time. Nevertheless, evidence is mounting that this component is now lingering in the soil. Different types of herbicides are at hand as an alternative to glyphosate – ones that lack such negative effects on the greater ecosystem. As an example, selective herbicides aimed at specific weeds rather than all plants they encounter, while non-residual alternatives break down faster rather than staying in the ground. If herbicides must be employed, making an informed decision and applying the best one (and only the required quantities) for each case is crucial.

It may appear as though farmers are at a crossroads independent of the path they go down. Yet options do exist. A main goal of regenerative agriculture is to better the resources utilized as opposed to depleting them. This places a significant premium on the health of soil, while sequestering CO_2 from the atmosphere, augmenting farm biodiversity and making water management better in the process. Limiting soil disturbance and chemical input application are key steps connected with the management style.

Agroforestry systems cover both traditional and modern land-use systems where trees are attended to in addition to crops and/or animal production systems in agricultural set ups. The mixture of trees, crops and livestock controls environmental risk, develops a permanent soil cover against erosion, reduces harm from flooding and acts as water storage, aiding crops and pastures.

The main concept in agroforestry is selecting the correct tree, in the correct location, for the correct use. Agroforestry methodologies may be designed to:

- care for soil, animals, crops, and homes from extreme climatic conditions
- increase water quality
- create jobs and income
- obtain goods (food, fiber, bioenergy, wood, floral and medicinal/ botanical products)
- protect pollinator habitat
- capture snow and stop it from piling on roads (living snow fences)
- act as a habitat for wildlife, and create corridors for their movement
- sequester carbon and lower other greenhouse gasses
- control odor
- create beautiful places

Agroforestry is a key method for healthy, long-term agricultural production. Agroforestry practices represent opportunities for landowners – both large and small – to diversify their production set up in order to achieve greater profitability and to control risk – all the while, increasing their environmental stewardship, and contributing to their neighbors and nearby communities and towns. Agroforestry may contribute to the provision of jobs and augment economic well-being in rural communities. Within a landscape, agroforestry can develop transition zones that help "reconnect" agriculture, people, and communities, creating a multi-functional and more sustainable landscape.

Within a watershed, agroforestry methodologies can represent a living ecosystem that connects headwater forests through agricultural lands to urban zones and on to the sea; bringing cleaner water for communities – both locally and downstream – and other public benefits, including those mentioned above.

Augmenting the quantity of agroforestry on farms, forests, ranches, cities, and community lands aid a community by:

1. helping with the creation of jobs and economic well-being in rural communities;
2. contributing to the sustainable production of food, fiber, and bioenergy; and
3. providing cleaner water, climate-resilient landscapes, increased carbon storage, and improved wildlife habitat.

Agroforestry systems are also part of "multi-cropping systems" that provide channels for farmers to limit risk – whether because of market volatility or climate conditions (Figure 4.3).

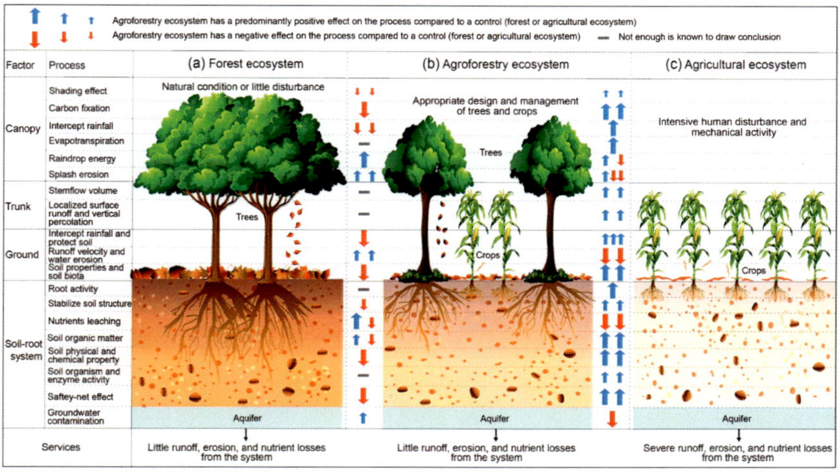

Figure 4.3. *Agroforestry traits and their potential impact on processes of reductions in water, soil and nutrient losses, and associated water contamination compared to the natural forest.*

Agroforestry is able to support farmers, ranchers, and woodland stakeholders in limiting the risk of sharp price increases or drops in a single commodity – because they will have a minimum of one alternative product to sell. Furthermore, if a flood or drought places an annual crop at risk, farmers will remain able to rely on their longer-term timber, fruit, or nut crops.

Agroforestry methodologies – those of incorporating trees with crops and/or livestock – have been employed by indigenous communities for centuries.

In addition, agroforestry practices are being championed as an answer to challenges that are often faced around the globe – achieving food security, lowering risk and improving income opportunities via diversification, promoting wildlife habitat, and providing key ecosystem services.

Agroforestry systems are dynamic and can be created to meet ranging economic, environmental and social targets. Although agroforestry may look different in tropic and temperate environments, it is of importance in both places.

Methodologies resembling agroforestry practices have been employed globally throughout history but were not referred to as agroforestry. Native Americans in North America had numerous complex management systems to produce their food, baskets, homes, clothes, fuel, medicine, and more. Similarly, early European settlers practiced a style of agriculture that incorporated trees, crops, and livestock as they worked with the endemic woodland vegetation and agriculture crops. Nevertheless, this approach mostly disappeared over the 20th century with development in separate research programs around agriculture and forestry that ended in current separate approaches and disciplines. There remain examples of integrated approaches being employed by farmers and forest managers, including Tribal members, and a rising interest in agroforestry from people around the globe.

CHAPTER 5

NUTRIENT ACQUISITION BY PLANTS

Contents

Nitrogen is the nutrient needed in the highest quantities by plants from soil, from which it is absorbed as ammonium (NH_4+) and nitrate ($NO_3^{-)}$. Nevertheless, NH_4^+ usually binds to clays and organic fractions in the soil, from which it is released over time, whereas NO_3^- can be leached from soil faster. The key sources of nitrogen include the decomposition of organic materials and from the fixation process by legumes. Phosphates are especially important for plant development and health in the early stages of vegetation growth, yet this element has low availability in both heavily acidic and heavily alkaline soils. In set vegetation, symbiotic links with root mycorrhiza play a significant role in plant acquisition of phosphorus. Potassium and secondary elements are needed in lower amounts by the majority of plants yet might be lacking in the early states of site reclamation. Nutritional challenges vary considerably between different forms of degraded land, despite some similar features.

Plant growth and development is impacted strongly by the combination and concentration of mineral nutrients found in the soil. Plants often have heavy difficulties in achieving an adequate supply of these nutrients to cover the demands of basic cellular processes because of their relative immobility. A lack of any one of them may translate into reduced plant productivity and/or fertility. Symptoms of nutrient deficiency often cover stunted growth, death of plant tissue, or yellowing of the leaves as a response to a reduced production of chlorophyll, a pigment needed for photosynthesis. Nutrient deficiency can have a strong effect on agriculture, producing lower crop yield or limiting plant quality. Nutrient deficiency can also translate into reduced overall biodiversity because plants serve as the producers that support the majority of food systems.

Alterations in the climate and atmosphere have strong impacts on plants, such as changes in the availability of key nutrients. In a world of continual climate change, it is paramount to understand the strategies that plants have developed to allow them to handle some of these challenges.

Two types of nutrients are thought of as key for plants: macronutrients and micronutrients. Macronutrients are the building blocks of essential cellular components such as proteins and nucleic acids; as the name hints at, they are needed in great quantities. Nitrogen, phosphorus, magnesium, and potassium are some of the most important macronutrients. Carbon, hydrogen, and oxygen are also considered macronutrients as they are needed in significant quantities to build the bigger organic molecules of the cell; nevertheless, they represent the non-mineral class of macronutrients. Micronutrients, including iron, zinc, manganese, and copper, are needed in lesser amounts. Micronutrients are often needed as cofactors for enzyme activity (Figure 5.1).

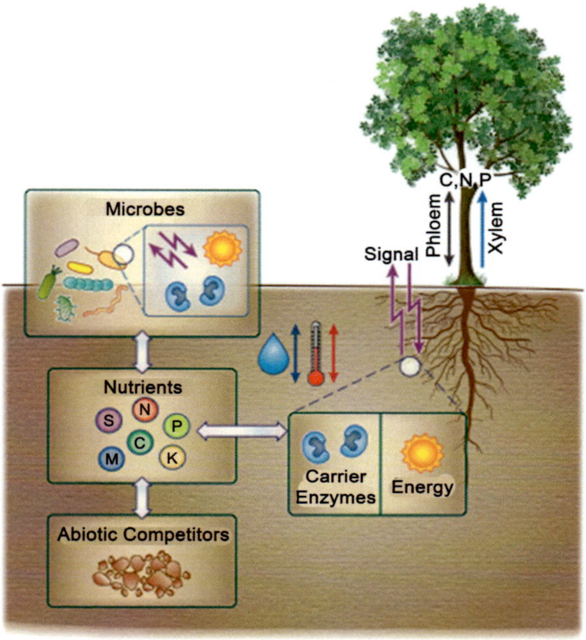

Figure 5.1. *Nutrient acquisition by plants.*

Mineral nutrients tend to be gathered from the soil via plant roots, yet several aspects can impact the efficiency of nutrient acquisition. To begin with, the chemistry and composition of some soils can make it difficult for plants to take up nutrients. The nutrients may not be there in certain soils, or may be present in forms that the plants cannot utilize. Soil characteristics such as water content, pH, and compaction can increase these problems.

Additionally, certain plants have mechanisms or structural features that represent advantages when growing in some types of nutrient-lacking soils. The majority of plants have developed nutrient uptake mechanisms that are adapted to their corresponding soils and are initiated in an attempt to surpass nutrient limitations. One of the most worldwide adaptations to nutrient-lacking soils is a change in the structure of the roots that may boost the overall surface area of the root to augment nutrient acquisition or may support elongation of the root system to reach new nutrient sources. These changes can conduct to a greater allocation of resources to overall root growth, in this way resulting in bigger root to shoot ratios in nutrient-limited plants (Lopez-Bucio et al., 2003).

Although nutrient deficiencies may represent significant threats to plant productivity, excess nutrients may become toxic, which is also problematic.

When some micronutrients build up to very elevated levels in plants, they add to the generation of reactive oxygen species (ROS), which can create impactful cellular damage. Some highly toxic elements such as lead and cadmium cannot be differentiated from essential nutrients by the nutrient uptake systems in the plant root, which means that in polluted soils, toxic elements may enter the food web by way of these nutrient uptake systems, causing lower uptake of the key nutrient and significantly lowered plant growth and quality. So as to keep nutrient homeostasis, plants must manage nutrient uptake and must address obstacles in the soil in addition to within the plant. Hence, plant species use numerous strategies for the movement and uptake of nutrients in addition to chelation, transport between the various cells and organs of the plant and storage to achieve whole-plant nutrient homeostasis. In the following sections a brief description of some examples of strategies used by plants to acquire nutrients from the soil are included.

5.1. PLANT ACQUISITION OF NUTRIENTS: SYMBIOSES WITH SOIL-BASED MICROORGANISMS

Nitrogen and phosphorus are two the elements thought of as most limiting to plant development and productivity since they are regularly present in minimal amounts locally or show up in a form that is unable to be utilized by the plant. Thus, the assessment of various plant species has covered the evolution of mutually beneficial symbiotic relationships with soil-borne microorganisms. As part of these relationships, both the host plant and the microorganism symbiont obtain valuable resources that they require for their self-productivity and survival as an outcome of the association.

Plants are tightly connected to microorganisms in rhizosphere soil, including those that give nutrients–fertilizer, alter the physical and chemical properties of soil, and the manipulation of hormone signaling (Hu et al., 2021; Liu et al., 2020). These are key in establishing plant health and can stop plant diseases from biotic or abiotic stresses through rhizodeposition (Dlamini et al., 2022). The rhizosphere area depicts a slot of microbial relationships that is set up upon the recognition of the plant root system. This area creates carbon fixed through photosynthesis and a collection of compounds covering microbial secondary metabolites, carbohydrates, amino acids, and vitamins, all of which have a function in strengthening soil fertility (Checcucci and Marchetti, 2020). It is an important source of energy input in plant–soil ecosystems that increases the diversity of plant–microbe interaction. The main result of which is the creation of unique plant root microbial groups. Hence, comprehending microbial variance in addition to their functions is a part of sustainable agriculture methodologies. Rhizosphere microorganisms is the term given to the varying types of microscopic lifeforms that live in the rhizosphere, the slim section of soil that is directly influenced by root secretions and linked soil microorganisms. This zone is known for an elevated level of biological activity because of the presence of substances produced by roots, including sugars, amino acids, organic acids, and numerous secondary metabolites. These compounds can be used as nutrients for microorganisms, fostering a plentiful and dynamic microbial population surrounding plant roots. The makeup of rhizosphere microorganisms is diverse and covers several important groups such as bacteria, fungi, protozoa, archaea, viruses (phages), and nematodes (Mendes et al., 2013; Saeed et al., 2021). These microorganisms are involved in a wide range of interactions with plant roots and each other, including symbiosis, competition, predation, and parasitism. Beneficial interactions, including those between mycorrhizal fungi or nitrogen-fixing bacteria and plant roots, strengthen plant nutrient acquisition and highlight tolerance. In the rhizosphere, bacteria stand out as one of the more regularly found and diverse groups of microorganisms.

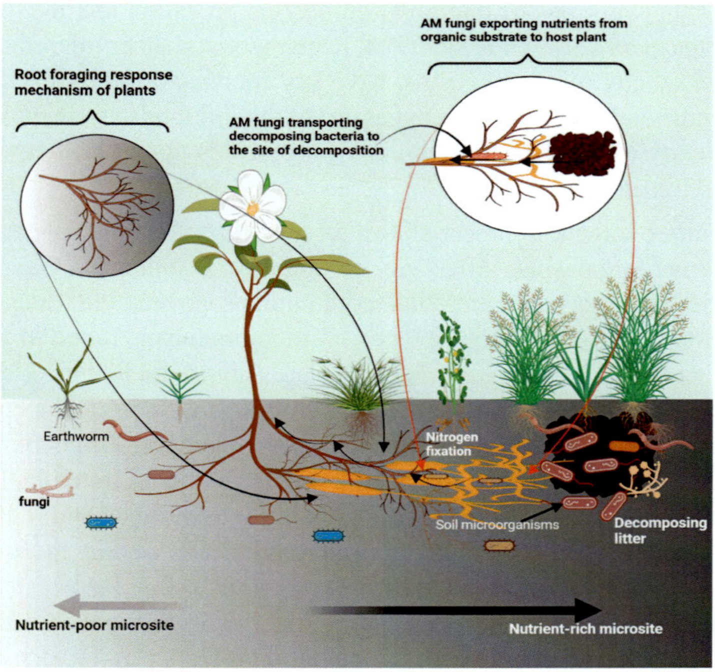

Figure 5.2. *The role of soil microorganisms in regulating the effects of soil nutrient heterogeneity.*

Source: Microorganisms 2022, 10(12), 2399; https://doi.org/10.3390/microorganisms10122399.

They are essential in nutrient cycling steps including nitrogen fixation and phosphorus and potassium solubilization, additionally they hold key functions in controlling diseases and supporting plant growth via the creation of numerous hormones and enzymes (Jacoby et al., 2017; Timofeeva et al., 2023) (Figure 5.2).

Fungi are key to this ecosystem, with symbiotic mycorrhizal fungi developing advantageous links with plant roots to strengthen water and nutrient uptake, while free-living fungi help in the breakdown of organic material and additional nutrient cycling (Wahab et al., 2023). Protozoa, the single-celled eukaryotes that feed on bacteria and other microorganisms, are essential for the management of microbial populations and nutrient cycling in the rhizosphere (Solomon et al., 022). Archaea, even though it resembles bacteria, are a genetically different group of microorganisms active in critical processes including ammonia oxidation, a key component of the nitrogen cycle (Wright et al., 2023). Viruses, especially phages that infect bacteria, show up in the rhizosphere as well, where they affect the workings of microbial groups via predation and the transfer of genes (Braga et al., 2020). Furthermore, nematodes and other microfauna, even though not exactly microorganisms, have key functions in the rhizosphere by adding to nutrient cycling and the dynamics of the soil food system, pointing out the complexity and interconnectedness of life in this unique soil environment (Jiang et al., 2018).

The unique soil environment allows for a myriad of ecological connections between microorganisms, affecting community taxonomy (Wang et al., 2020). The taxonomy of microbial communities covers the categorization and understanding of the varying types of microorganisms found in a specific environment based on their evolutionary relationships and features (Leff et al., 2018). This classification covers from broad groups to more niche categories, following a hierarchical setup characterized by domain, kingdom, phylum, class, order, family, genus, and species. In the world of microbial groups, taxonomy provides a framework for organizing this variance and highlighting the roles and connections of various microorganisms within the group. As an example, bacterial phyla including *Bacteroidetes* and *Proteobacteria* are boosted in the rhizosphere, whereas phyla like *Chloroflexi* and *Acidobacteria* are more often found in bulk soil. This setup is affected by factors including plant type, soil conditions, and agricultural practices (Ling et al., 2022; Vorholt et al., 2017). Recent research suggested that protists in the rhizosphere may defend plants by feeding on other microorganisms and moving the taxonomic and functional composition of microorganisms of the rhizosphere (Xiong et al., 2020). Similarly, the study of Zhang et al. (2021) suggested that rhizospheric protists have the ability to influence bacterial and fungal groups to influence the rhizospheric co-occurrence relationships of soybean plants. In addition, identifying a microorganism's taxonomic position may hint at its physiological properties (Zhu et al., 2015).

The physiological characteristics of microorganisms cover a wide range of functional features that allow their development, survival, and behavior in changing environments. These characteristics cover a variety of metabolic pathways, enabling them to use different forms of carbon and energy via procedures including aerobic and anaerobic respiration, fermentation, photosynthesis, chemolithotrophy, and methanogenesis (Wani et al., 2022; Counts et al., 2017). Microorganisms additionally count with set mechanisms for nutrient absorption and assimilation, key for processing carbon, nitrogen, phosphorus, sulfur, and other trace elements and vitamins (Kiri et al., 2023). Their development is affected by optimal conditions linked to temperature, pH, and salinity, with adaptations that allow some to thrive in extreme conditions, resisting desiccation, high salinity, and extreme temperatures (Kochhar et al., 2022).

Communication is another paramount physiological characteristic, with numerous microorganisms making use of chemical signals for quorum sensing to manage collective behaviors including the formation of biofilm (Preda et al., 2019). Movement is made possible by structures such as flagella, cilia, or pili, enabling displacement in response to environmental signals (Aroney et al., 2021). Fungi on the other hand have evolved unique strategies for moving and colonizing new environments by the dispersion of spores and hyphy development (Coleine at al., 2022). Mechanisms for the storage of energy permit the accumulation of resources in forms including glycogen, polyphosphate, or sulfur globules. To protect from threats, microorganisms can develop antibiotics, toxins, and form biofilms, and have evolved different types of resistance mechanisms. Finally, the cellular structure, including cell wall composition and the use of specific organelles or appendages, have a key function in their physiological features.

Beneficial rhizosphere fungal and bacterial groups are well-known as plant growth promoting microorganisms (PGPM). These microorganisms are able to colonize plant roots and help their hosts by synthesizing phytohormones, enhancing soil nutrients, fighting pathogens infection, and developing a tolerance to biotic and abiotic stress, making way for a lowering in the utilization of pesticides and synthetic fertilizers in crop (Adeleke at al., 2021; Pereira et al., 2021). Host plants give rhizo-deposit nutrients, boundary cells, mucilage, lignin, and cellulose as food sources for microorganisms via their roots. These are a dynamic collection of organic compounds synthesized by plant roots such as sugars, amino acids, organic acids, fatty acids, vitamins, nucleotides, and various secondary metabolites (Vives-Peris et al., 2020). PGPM are able to break down boundary cells such as those composed of cellulose in plant roots by way of a combination of enzymatic degradation, invasion strategies, and symbiotic or pathogenic connections (Cosgrove 2023; Yi et al., 2022). They aid in the breakdown of plant biomass, releasing nutrients that can be used once again by plants and other organisms. PGPM answer to plants by storing nutrients via solubilization, phosphorus and potassium solubilization, and nitrogen fixation and the creation of plant hormones in the rhizosphere, which support plant health to promote plant development (Olanrewaju and Babalola, 2022; Upadhyay et al., 2022).

In terrestrial ecosystems, plants are connected with a diversity of soil microbial groups that set up interdependent relationships (Mitter et al., 2016), which foster plant group productivity (van der Heijden et al., 2008), belowground biodiversity, and ecosystem multifunctionality (Harris et al., 2009; Johnson et al., 2005; Wagg et al., 2014). These connections are essential for numerous reasons (Chandra, 2019; Read and Perez-Moreno, 2003), such as the nutrient uptake of plants from heterogeneously distributed microsites. The reactions of plants to spatially, heterogeneously distributed nutrients in the soil need a specific physiological strategy often known as the root foraging process. An example of this is the spread of roots in nutrient-rich microhabitats (Cahill et al., 2010; Hodge et al., 2004) and microbially controlled mechanisms through a plant–microbe symbiotic connection to guarantee the effective uptake of soil nutrients (Hestrin et al., 2019). As an example, arbuscular mycorrhizal fungi are key to these plant–soil microbe relationships (Kawaguchi and Minamisawa, 2010). Although both soil microorganisms and soil nutrient heterogeneity are likely to impact plant performance, community productivity, and competitive interactions (van der Heijden et al., 2008), not much is known about the potential effects of soil microorganisms on the impact of soil nutrient heterogeneity on plant development (Du et al., 2009; Adomako et al., 2021).

Soil nutrient heterogeneity is the term used to describe the variability in the distribution of the available soil nutrients within the soil matrix in a given microhabitat (Hodge, 2005; Stark, 1994). The varying distribution of nutrient resources at hand may sprout from different sources, including uneven distribution and breakdown of litter in the soil (Cai et al., 2019; Loydi et al., 2013), variations in the parent material during weathering processes, topography, climate, and differences in the availability of microorganisms (Cai et al., 2019; Loydi et al., 2013). Spatial heterogeneity of soil nutrients is omnipresent in the ecosystem and plays a fundamental role in the development of the individual plant (Adomako et al., 2021; Dong et al., 2015; Zhou et al., 2012), population structure (Day et al., 2003), group output (Birch and Hutchings, 1994; Adomako et al., 2020), intraspecific and interspecific interaction, and species coexistence (Zhou et al., 2012; Xue et al., 2018; de Kroon et al., 2012; Zhang et al., 2022). One mechanism behind such an effect on plants is the root foraging response from the roots in nutrient-rich microsites (Hodge et al., 2004; Hutchings and Kroon, 1994). Nevertheless, the size or scale of the heterogeneity (i.e., patch scale) (Jackson and Caldwell, 1993) and the varying nutrient level (i.e., patch contrast) are the two main features of soil nutrient heterogeneity that may affect plant growth (Wang and Alpert, 2016).

A frequent reason is that the reply of plants to soil nutrient heterogeneity through the foraging mechanism are strongly reliant on both patch scale and patch contrast (Hutchings and Wijesinghe, 2008). Existing research has demonstrated that a plant is able to answer to soil nutrient heterogeneity at one scale but may be unresponsive at another (Cahill et al., 2010; Bliss et al., 2002; Chen et al., 2019). Hence, the patch size and contrast of soil nutrients may dictate the effect of spatial heterogeneity on plant development (Fransen et al., 1998; Qian et al., 2014). However, the activities of soil microorganisms can affect the distribution of soil nutrients regardless of the patch size or patch contrast (Adomako et al., 2021; Reynolds and Haubensak, 2009), thus altering the impact of soil nutrient heterogeneity on plant life.

The soil microbial groups make up a diverse community of microorganisms whose activities may benefit or affect the development and productivity of plants (van der Heijden et al., 2008; Chandra, 2019). As such, soil microorganisms directly impact plant development by creating a mutual (symbiotic) or pathogenic connection with the roots and, via the free-living microorganisms (non-symbiotic) that are indirectly able of switching the speed of nutrient supply to plants (van der Heijden et al., 2008). Common among these root-associated microorganisms are arbuscular mycorrhizal fungi (AMF), which have the ability to supply limiting nutrients including soil phosphorus (P) to its host in exchange for carbon (Georgiou et al., 2017).

This AMF-plant connection is the oldest and most abundant connection in the terrestrial ecosystem (Hodge et al., 2000; Johnson, 2010). Additionally, the N-fixing bacteria give the highest quantity of soil nitrogen for plant group productivity in the majority of ecosystems, especially in plant groups dominated by legumes (van der Heijden et al., 2008; Pajares and Bohannan, 2016). This nitrogen represents approximately 20% of the total nitrogen required by plants every year (Cleveland et al., 1999; Van Der Heijden et al., 2006). Indirectly, there is a significant amount of microorganisms that are nitrogen-fixers but are not in a direct symbiotic relationship with any vascular plant, for example free-living N_2 fixers, cyanobacteria that fix N2 via its symbiotic association with lichens and bryophytes (Cleveland et al., 1999). In addition, some filamentous actinomycetes, including *Frankia*, are known to fix N2 as free-living or when in a symbiotic relationship with non-luminous vascular plants (Cleveland et al., 1999; Benson and Silvester, 1993). Regardless of these pieces of accumulating evidence that soil microorganisms may affect the growth of plants, knowledge of how these belowground groups manage plant nutrient acquisition in patchily distributed soil nutrients in the terrestrial ecosystem is still minimal.

5.2. Nitrogen Fixation

Even though nitrogen is the most abundant gaseous element in the atmosphere, plants are not capable of using the element in this form (N_2) and may experience a deficiency of nitrogen in some soils with a low nitrogen content. Because nitrogen is a basic component of both proteins and nucleic acids, nitrogen deficiency creates significant limitations to plant productivity. In an agricultural setting, nitrogen deficiency can be solved by adding nitrogen-rich fertilizers to augment the availability of nutrients and thus increase crop yield. Nevertheless, this practice can be harmful as excess nutrients tend to end up in groundwater, creating eutrophication and subsequent oxygen deprivation of connected aquatic ecosystems (Figure 5.3).

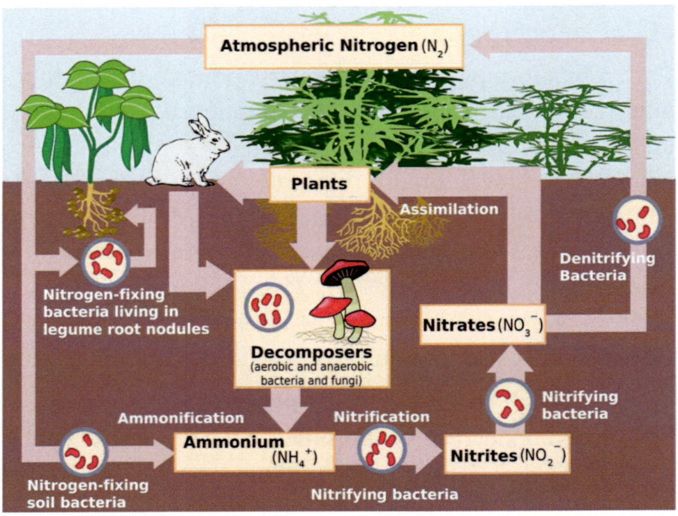

Figure 5.3. *Atmospheric nitrogen.*

Plants are capable of directly acquiring nitrate and ammonium from the soil. Nevertheless, when these sources of nitrogen are lacking, certain types of plants from the family *Fabaceae* (legumes) comense symbiotic connections with a group of bacteria called *Rhizobia*, which are able to fix nitrogen. These relationships are relatively specific and need the host plant and the microbe to recognize each other via chemical signals. The interaction starts when the plant releases compounds called flavanoids into the soil that attract the bacteria to the root. As a reply, the bacteria release compounds called Nod Factors (NF) that create local changes in the structure of the root and root hairs. Specifically, the root hair curls sharply to envelop the bacteria in a small pocket. The plant cell wall is disintegrated and the plant cell membrane invaginates and creates a tunnel called an infection thread that grows to the cells of the root cortex. The bacteria are thus wrapped in a plant derived membrane as they morph into structures called bacteroids. These structures are allowed to enter the cytoplasm of cortical cells where they turn atmospheric nitrogen to ammonia, a form that can be utilized by the plants. In exchange, the bacteroids receive photosynthetically derived carbohydrates to use for the production of energy (Limpens and Bisseling, 2003; Ferguson et al., 2010).

CHAPTER 6

SOIL COMPOSITION INFLUENCES NUTRIENT AVAILABILITY TO PLANTS

Contents

Although the majority of plants are autotrophs and are able to create their own sugars from carbon dioxide and water, they still need certain ions and minerals found in the soil. This process is controlled by root hairs, which are extensions of the root epidermal tissue that increase the surface area of the root, significantly adding to the absorption of water and minerals. Nevertheless, the properties of the soil can directly impact the availability of the specific ions present in the soil. The most essential soil property that influences ion availability is the presence (or absence) of clay:

Clay is negatively charged, and therefore any positive ions (cations) that are found in clay-rich soils will stay tightly bound to the clay particles. This tight connection with clay particles stops the cations from being washed away by heavy rains, but it also stops the cations from being easily taken up by plant root hairs.

On the other hand, the negatively charged anions are dissolved quite simply in soil water and hence readily available to plant root hairs; nevertheless, they are also quite simply washed away by rainwater.

As an outcome from the factors noted above, the presence of significant quantities of clay particles in soil act as a trade-off for plants: clay particles stop leaching of cations from the soil by rainwater, yet they also stop absorption of the cations by the plant.

Other soil compositions carry different forms of tradeoffs for plant roots:

Sandy soil are made up of loosely packed soil particles, translating into lots of air pockets that promote root penetration and respiration; nevertheless, the loose packing also implies that water drains away quite fast– usually carrying numerous nutrients away with it.

Clay soil holds water well as the water molecules stay connected with the charged clay surfaces; nevertheless, clay particles packed tightly together meaning that there is lower amounts of air available in the soil and posing larger challenges for plant roots to infiltrate the dense soil.

Big quantities of organic matter in soil signify a near-ideal environment for plant roots, with large concentrations of nutrients, elevated water retention, and loose soil packing with numerous air pockets.

The information described above is organized in Table 6.1.

Table 6.1. Soil Composition and Nutrient Uptake

Soil composition	Water availability	Nutrient availability	Oxygen availability	Root penetration ability
Sand	Low: water drains out	Low: poor capacity for cation exchange; anions leach out	High: many air-containing spaces	High: large particles do not pack tightly
Clay	High: water clings to charged surface of clay particles	High: large capacity for cation exchange; anions remain in solution	Low: few air-containing spaces	Low: small particles pack tightly
Organic matter	High: water clings to charged surface of clay particles	High: ready source of nutrients, large capacity for cation exchange; anions remain in solution	High: many air-containing spaces	High: large particles do not pack tightly

6.1. PROCESSES AND STRUCTURES THAT MOVE MOLECULES INTO AND OUT OF CELLS

Prior to the discussion on how plants take up nutrients from the soil, several structures and process that control movement of molecules across cell membranes must be described:

1. Diffusion takes place when molecules move from an area of high concentration to an area of low concentration. Diffusion does not need energy since the molecules move "down" their concentration gradient, meaning from areas of high to low concentration.
2. Facilitated diffusion refers to diffusion that happens through a protein channel located in a cell membrane. Each protein channel is highly specific for its given molecule. Facilitated diffusion is required for molecules that are too large or too polar to cross the nonpolar cell membrane.
3. Active transport means movement of a molecule *against* its concentration gradient. Active transport always needs the utilization of energy, usually (but not always) from ATP.
4. Proton pumps a complex of proteins that utilize energy in the form of ATP to "pump" protons from one side of a membrane to another. This sets off an electrochemical gradient, with an elevated concentration of positively-charged protons on one side of a plasma membrane. This electrochemical gradient is then able to be employed as a source of energy to mobilize other molecules against their concentration gradients via co-transporters.

5. Co-transport refers to the movement of two molecules simultaneously: one molecule is transported along ("down") its concentration gradient, which creates an energy expenditure that is utilized to transport the other molecule *against* its concentration gradient. Co-transport occurs through protein channels called co-transporters. As with all protein channels, co-transporters are highly specific for the given molecules that they can transport.

6.2. PLANT UPTAKE OF MINERAL NUTRIENTS FROM THE SOIL

This process depends on proton pumps, cation channels, and anion cotransporter channels present in the membranes of the root hairs as follows:

1. The epidermal tissue of root hairs is covered by proton pumps (H^+ ATPases), which utilize ATP as a source of energy to pump protons out of the cells and into the soils, against their electrochemical gradient. These proton pumps develop a strong electrochemical gradient with an elevated concentration of protons and a strong positive charge outside of the cell, and a minimal concentration of protons and relatively negative charge inside of the cell. These protons pumped into the soil by the proton pumps create two direct outcomes:
2. The positively-charged protons link to the negatively-charged clay particles in the soil, freeing the cations from the clay in a process known as cation exchange. The cations then diffuse down their electrochemical gradient into the root hairs via cation channels. Because the soil environment is strongly positively charged, it is energetically favorable for cations to move into the root hairs and out of the soil environment.
3. The elevated concentration of protons in the soil develops a strong electrochemical gradient that facilitates transport of protons *back* into the root hairs. Plants use co-transport of protons *down* their concentration gradient as the source of energy to additionally move anions *against* their concentration gradient into the root hairs. This process happens via anion cotransporter channels. Because the soil environment is strongly positively charged, it is energetically unfavorable for anions to move out of the soil, but highly energetically favorable for protons to abandon the soil.

6.2.1. Mycorrhizal Interactions with Plants

Apart from the symbiotic relationships with bacteria, plants are also able to participate in symbiotic relationships with fungal organisms. Modern day approximations of the frequency of plant-mycorrhizal relationships proposes that an estimated 80% of all plants set up some form of mycorrhizal symbiosis, and numerous investigations show that these relationships are millions of years old (Karandashov and Bucher, 2005; Vance, 2001). There are various types of mycorrhiza, ranging in structural morphology, the method of colonizing plant tissue, and the host plants colonized. Nevertheless, there are two main types that are usually considered as the most often found and thus, the most ecologically impactful. The endomycorrhizae are those fungi that set up associations with host plants by entering the cell wall of cortical cells in the plant roots. On the other hand, ectomycorrhizae create a large network of hyphae across cortical cells but do not truly penetrate the cells.

The most often found endomycorrhizal connection takes place between arbuscular mycorrhizal fungi (AMF; also called Vesicular-Arbuscular Mycorrhiza or VAM) and a diversity of species of grasses, herbs, trees and shrubs. When phosphate is present in the soil, plants can get it directly through root phosphate transporters. Nevertheless, under minimal phosphate conditions, plants turn dependent on connections with mycorrhizal fungi for phosphorus acquisition. Mycorrhizal spores available in the soil are germinated by compounds liberated from the plant. Hyphae extend from the germinating spore and enter the epidermis of the plant root. Inside the root, the hyphae branch and penetrate cortical cells, where heavily branched structures named arbuscules evolve. Outside, hyphae extend into the soil beyond the zone accessible to the root. This type of symbiosis supports plant phosphorus absorption from the soil by expanding the root's absorptive surface area. Because plants absorb phosphorus at a much faster rate than phosphorus diffuses into the soil around the root, a phosphorus depletion zone is rapidly established, controlling uptake of phosphorus by the plant. Nevertheless, AMF hyphae create a bridge between the internal root environment and the area outside this depletion zone allowing the plant to get significantly more phosphorus via its symbiotic partner than it could alone (reviewed by Karandashov and Bucher 2005).

Ectomycorrhizal fungi (EcM) establish associations with various tree species (birch, oak, spruce, pine, fir), turning these microorganisms into a key component of the biodiversity of the forest ecosystems, especially temperate and boreal forests in the northern hemisphere. Even though the molecular mechanisms behind the initiation of EcM-plant interactions remain not well defined, it is suggested that signaling molecules are liberated by both symbiotic partners, creating fungal hyphae growth towards compatible plant roots. The EcM creates a large hyphal sheath, or mantle, surrounding the root cap and a strong net of hyphae that expands into the root itself and goes around the cells within the root cortex. Hyphae from the mantle expand into the surrounding soil, giving availability and accumulation of nutrients such as phosphorus and nitrogen for the plant. This is especially impactful for trees growing in forest soils since a big part of the nutrients accessible are found in the litter layer and are not available to the roots of these trees. The EcM creates enzymes that are able to digest this organic material and additionally have the capacity to move the nutrients to the Hartig net, making them accessible to the plant. In this way, the interaction between trees and EcM is not only useful, but often, is paramount (Nehls et al., 2008).

CHAPTER 7

UNDERSTANDING NUTRIENT CONTENT IN FOOD

Contents

When it comes to understanding the nutrient content in food, it is essential to consider the role that soil quality plays in determining the nutritional value of crops. Soil quality refers to the physical, chemical, and biological properties of the soil that influence plant growth and development. Factors such as soil pH levels, organic matter content, and soil microbiology all play a crucial role in determining the nutrient content of food.

Research has shown that soil pH levels can have a significant impact on the nutrient content of crops. Soil that is too acidic or too alkaline can limit the availability of essential nutrients to plants, resulting in lower nutrient content in the harvested crop. It is essential for farmers to regularly test and monitor soil pH levels to ensure optimal nutrient availability for plants.

7.1 VARIATIONS IN FARMING PRACTICES

Organic farming practices have also been shown to have a positive impact on soil quality and nutrient content in food. Organic farming methods focus on building healthy soils through the use of compost, cover crops, and crop rotations. These practices help to improve soil structure, increase soil organic matter, and enhance soil microbiology, all of which contribute to higher nutrient levels in crops.

In contrast, conventional farming practices, which rely heavily on synthetic fertilizers and pesticides, can degrade soil quality over time. Soil erosion, caused by intensive tillage and monocropping, can also lead to nutrient depletion in crops. Sustainable agriculture practices, such as no-till farming and agroforestry, can help to protect soil health and maintain nutrient content in food.

By understanding the relationship between soil health, plant health, and nutrient content in food products, we can work towards promoting sustainable agriculture practices that prioritize soil quality and nutrient-rich food production. Soil testing methods can be used to assess nutrient levels in food crops and guide farmers in making informed decisions about soil management practices. By focusing on improving soil quality, we can ensure the nutrient content of our food remains high and support the health and well-being of consumers (Hargrove, 2003).

7.2 HISTORICAL CHANGES IN NUTRIENT CONTENT OF FOOD

The nutritional values of some commonly consumed vegetables, ranging from asparagus to spinach, have decreased strongly since 1950. A 2004 investigation from the United States found key nutrients in certain garden crops are approximately 38% below where they were in the mid-20th Century

(Davis et al., 2004). On average, from the 43 vegetables analyzed, calcium content dropped by 16%, iron by 15% and phosphorus by 9%. Vitamins such as riboflavin and ascorbic acid both reduced strongly, while there were small reductions in the levels of protein. Comparable decreases have been seen in the nutrients found in wheat (Fan et al., 2008).

Inspired by food shortages following World War Two, researchers developed new high-yield varieties of crops and breeds of livestock, in tandem with synthetic fertilizers, pesticides and herbicides, to increase food production. In addition to irrigation improvements and the onset of affordable tractors, crop productivity went up dramatically. The average global cereal yield increased by 175% from 1961 to 2014 (Ritchie, 2017), with wheat, as an example, increasing from an average yield of 1.1 tons per hectare to 3.4 tons per hectare (Ritchie et al., 2022) in a similar time span.

While yields increased, nutrient levels in certain crops went down, placing intensive farming techniques under watch. Is it possible, as it has been suggested (Marles, 2017), the result of the greater use of artificial pesticides, fertilizers and other chemicals impacting the delicate balance of soil life, the health of crop plants, and thus impacting the quality of the food?

A 170-year study (Blair et al., 2017) into wheat produced utilizing various farming techniques in the United Kingdom proposes there is more taking place. "The Broadbalk experiment is one of the oldest continuous agronomic experiments in the world. Started in 1843, it has been comparing the effect of inorganic [artificial] fertilizers and organic manures on winter wheat. It has specifically examined the levels of iron and zinc in wheat grown under different farming methods."

To start with, their findings demonstrate that it is not missing micronutrients in the soil that is creating the lower nutrients in the crop. Those that are bioavailable, meaning, in a form that the plant can absorb, do not change with intensive farming methods.

If the soil is in the same condition as it was, what else is happening? Have the plants themselves changed?

In the 1950s, an American scientist by the name of Norman Borlaug researching in Mexico developed "semi-dwarf" types of disease resistant-wheat. By lowering the stalk height by 20%, the plants were much less likely to topple over – a problem referred to as "lodging" – which lowered their productivity and turned diseases more likely to remain, in addition to rendering mechanical harvesting much less effective (Jobson et al., 2017).

An added plus of the discovery of those dwarfing genes was that instead of putting the energy into developing a longer stalk, the plant put it into the spike [the ear, where the wheat grains grow. Rather, the smaller plant feeds carbohydrates into the grain, elevating the quantity of grain per plant. The plant managed this by favoring an engrossed wheat grain endosperm, which the seed makes use of to feed the developing plant embryo similarly to how an egg yolk feeds a growing chick. This is dense with carbohydrate in the form of starch – the main component of flour.

Obtaining larger and more plentiful grains was a substantial improvement when populations in developing countries were expanding quickly and famine was a concern. Nevertheless, an unexpected side effect was that even though the wheat produced more grain per plant, nutrient levels did not go up in the same way.

In this case, even though the nutrients stay at the same amounts in a single wheat kernel, the starch is increased by approximately two or three-fold. This means that once the wheat is turned into flour one obtains a dilution effect. The ratio of carbohydrates to nutrients is lowered.

Even though carbohydrates are essential for human health – they provide the energy that supports movement and daily functions – the food selection must also be a source of protein, minerals and vitamins which are essential for growth and biochemical processes in the body. Selenium, for example, is required in processes where DNA is made, zinc supports the body's immune system to work properly, while magnesium keeps nerve, muscle and heart functions, and helps bones stay strong.

While the Green Revolution aimed to lower world hunger, today the world has a global food system that in some areas has been created to provide calories and cosmetic desirability but not always nutrition. This is adding to a phenomenon known as hidden hunger, where people feel satisfied but may not be healthy, as their food is rich in energy (calories) but deficient in nutrients. While it might sound counterintuitive, it is possible for obese individuals to be nutrient-deficient (Via et al., 2012).

7.3 THE IMPACT OF SOIL QUALITY

Even though certain researchers consider the decrease in nutrient levels in the food over the decades is too minimal to be significant (Marles et al., 2017) in comparison to the increase in food availability given by improved crops, soil health is still considered to have a key relationship to the nutritional quality of food. A study in the United States has been looking at vegetables grown under different farming techniques to better comprehend exactly what this connection is.

"The vegetable systems trial started in 2016 and is a side-by-side comparison of crops grown in soils managed with intensive practices and regenerative organic tillage practices," says soil scientist Gladis Zinati of the Rodale Institute in Pennsylvania.

The goal of their research, which is set up to run indefinitely, is to connect farming practices and soil health to crop nutrient density (or the quantity of nutrients a food contains in comparison to the amount of calories) and the health of people.

Zinati's investigation proposes that the more fungi and microbes that are active in the soil, the better able this soil is to make nutrients available for plants and thus the diet. Meaning, soil equipped with fungi and bacteria is better able to metabolize nutrients into a form that can be more easily absorbed by the plants.

Soil is composed of four elements in diverse combinations:

1. minerals in the form of rock particles
2. organic matter (plant, fungal and animal materials including microbes and microscopic worms, either dead or alive)
3. air
4. water

Yet the key factor is the way in which these elements interact. A teaspoon of soil contains more microbes than there are people on the planet – approximately billions, and as many as 10,000 individual species (UKCEH, 2024).

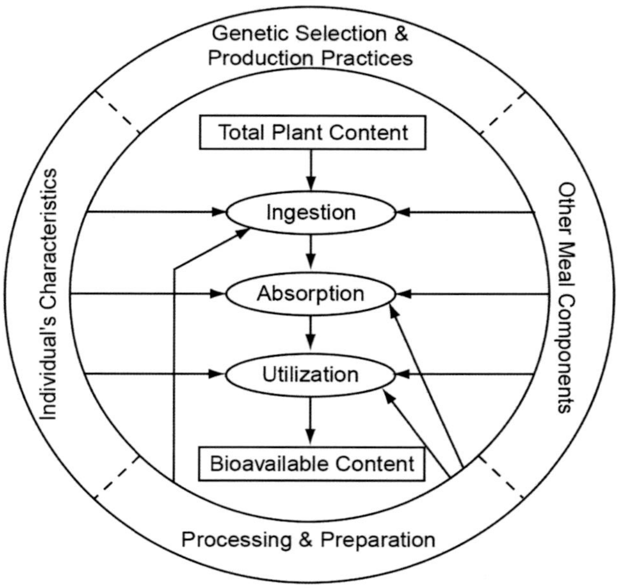

Figure 7.1. *Factors affecting micronutrient bioavailability to humans from plant food sources.*

Mixed in with all that mind-boggling activity is a network of fungal filaments referred to as mycorrhiza, a symbiotic relationship between plants and microbes that function as extensions to their roots. There is an ongoing dance of nutrients with an exchange from soil to plant, fueled by complex biochemical pathways, fungal superhighways and elegantly named root exudates, which stimulates or represses a large number of biological activity in the soil (Figure 7.1).

Specialized fungi found in the soil (an inoculant) effectively expand the plant root system with mycelium – a web of long microscopic filaments called hyphae. They work in a symbiotic relationship with the plant, liberating nutrients from deep in the soil, in a form that the plant can take up. In food production, an inoculant is a beneficiary microorganism applied to either the soil or the plant in order to improve productivity and crop health.

As an example of the use of inoculants, consider a United States farmer who farms approximately 32 sq km of corn and soybeans in Ohio and has been trialing the use of inoculants. In addition to observing an increase in yield through the release of nutrients from the soil, a reduction of costs from the use of chemical inputs can be achieved.

By making use of naturally occurring inoculants, the focus is on soil health. Thus, once the soil health aspect is secured, it will translate into plant health. Although the use of synthetic fertilizers is still applicable, their quantity is becoming less and less, approximately 25% lower than originally used. The idea is to break apart more of what is already in the soil, rather than constantly adding more to the soil. In this case, less is more.

In addition to the fungi that have become highly adapted and are improving food productivity and the exchange of nutrients – plants that have evolved to survive demanding conditions are also proving to be beneficial.

In Kenya for example, livestock are an essential part of the economy, representing approximately 12% of the national GDP. In the smallholder combined crop-livestock system, livestock play a key function in food and nutritional security, income generation and as a source of fertilizer for soil used for crops.

In Kenya, dairy farming is particularly of value. This country is one of the most important dairy producers in Africa. Nevertheless, low quality feeds and seasonal scarcity cap productivity, with the average yield at approximately eight liters per cow each day in comparison to the 25–50 liters per cow each day in other regions of the world.

In Kenya, the majority of smallholder dairy farmers in the country possess only two to five cows. Thus, any improvement in productivity can have a meaningful impact on their lives. More than 90% of smallholder dairy farmers depend on Napier grass, which is employed in a 'cut-and-carry' feeding system, in which the grass is harvested and delivered to the animals. Nevertheless, Napier has turned susceptible to pests and diseases, which lower biomass yield strongly. There was a need to obtain other viable and sustainable options that farmers can rely on for their dairy livestock.

A type of grass called Brachiaria has entered the market in South America, Australia and Asia where it has altered the beef and dairy industry in those areas. Surprisingly, *Brachiaria* grass is originally from Africa, yet it has not been exploited there for cattle feed in a similar manner – until now.

The effect has been impactful. Milk yields increased, the health of the cows got better, crude protein levels increased in comparison to Napier grass and it is less fibrous and more digestible, meaning animals feeding on it produce lower amounts of greenhouse gasses. Because of its massive root system, it is also able to sequester more carbon into the soil than other local pastures.

The "super-grass's" success is uncomplicated – it has adapted to drought and low fertility acidic soils by developing a large, extensive root system, thus has the ability to absorb a higher amount of nutrients from deep in the soil.

Whether it is natural or human selection, farming techniques or the weather, the nutritional value of food is impacted by many factors. Ensuring that the best version of the foods grown is achieved needs an in-depth consideration of the network of nutrients that flows around us.

What would be required is a food production system which tracks nutrition in food and makes it universally comparable, and a commercial model that values nutrition above everything else. How that may be achieved remains to be established.

Food producers must be paid for effective nutrient yield, not simply for a mass of produce. Currently, the model of being paid per ton of grain is incompatible with a human health perspective.

There are many components in the connection between nutrition and farming, and a vast amount is not yet fully understood. As is often the case, more research is needed, but with more than two billion people worldwide impacted by micronutrient deficiencies, so much good could come from following the trail of nutrients.

CHAPTER 8

LINK BETWEEN SOIL QUALITY AND FOOD NUTRIENT CONTENT

Contents

The link between soil quality and food nutrient content is a crucial aspect of agriculture that has far-reaching implications for food scientists, university students, researchers, and policy makers alike. Soil quality plays a significant role in determining the nutrient content of food, as it directly impacts the availability of essential minerals and nutrients for plant uptake.

8.1 PH LEVELS

One key factor that influences the nutrient content of food is soil pH levels. Different crops have specific pH requirements for optimal growth and nutrient uptake. Soil pH levels can affect the availability of nutrients such as phosphorus, potassium, and calcium, which are essential for plant health and the nutritional quality of food.

Organic farming practices are known to promote soil health and enhance nutrient content in food. Organic farming avoids the use of synthetic chemicals and pesticides, which can degrade soil quality over time. Conventional farming practices, on the other hand, may rely heavily on chemical inputs that can negatively impact soil health and nutrient availability.

8.2 SOIL EROSION

Soil erosion is another critical issue that can lead to a loss of nutrients in crops. Erosion can deplete the topsoil of essential minerals and organic matter, resulting in decreased nutrient content in food products. Sustainable agriculture practices, such as conservation tillage and cover cropping, can help prevent soil erosion and maintain soil quality for optimal nutrient uptake by plants.

Soil microbiology plays a vital role in nutrient availability for plants. Soil microbes break down organic matter and release nutrients in a form that plants can absorb. The use of compost and other soil amendments can enhance soil microbiology and improve nutrient content in food crops.

Overall, understanding the relationship between soil quality, plant health, and nutrient content in food products is essential for promoting sustainable agriculture practices and ensuring a nutrient-rich harvest for future generations. Soil testing methods can help assess nutrient levels in food crops and guide farmers in making informed decisions to improve soil quality and enhance the nutritional value of their produce.

CHAPTER 9

CAUSES AND CONSEQUENCES OF SOIL EROSION

Contents

Erosion of soil is a deterioration of land due to the removal of its particles. It consists of three basic stages: *dislodgement, transportation, and sedimentation*. Their speed depends on the soil type, aggregation, infiltration, and ground cover. For example, well-aggregated fields are less vulnerable, and bare lands are the easiest to be destroyed. Slopes require additional protection from water erosion during rain showers, which can be addressed with cover crops, perennial grasses, or terrace farming.

Soil erosion is a significant problem that can have serious consequences on the nutrient content of food crops. There are various causes of soil erosion, including deforestation, overgrazing, improper agricultural practices, and extreme weather events. When soil erosion occurs, valuable topsoil, which is rich in nutrients essential for plant growth, is lost. This can lead to decreased soil fertility and reduced nutrient availability for crops.

The consequences of soil erosion on food nutrient content can be far-reaching. As topsoil is lost, essential nutrients such as nitrogen, phosphorus, and potassium are also lost. This can result in crops that are deficient in these nutrients, leading to lower yields and poorer quality produce. In turn, this can have negative impacts on human health, as nutrient-deficient crops may not provide the necessary vitamins and minerals for a healthy diet (Lal, 2003).

To address soil erosion and its impact on nutrient content in crops, it is important to implement sustainable agricultural practices. This includes practices such as contour plowing, cover cropping, and crop rotation, which help to prevent soil erosion and maintain soil health. Additionally, the use of compost and other soil amendments can help to improve soil fertility and nutrient content in crops.

It is essential for food scientists, university students, researchers, and policy makers to understand the relationship between soil erosion, soil quality, and food nutrient content. By implementing sustainable agriculture practices and taking steps to protect soil health, we can ensure that our food supply remains nutrient-rich and provides the essential vitamins and minerals needed for a healthy diet.

9.1. NATURALLY OCCURRING CAUSES OF SOIL EROSION

- **Strong Wind Gusts:** Heavy winds take away dry tiny earth particles, which is a classic problem in semi-arid regions making way for desertification.
- **Climate Change:** Unexpected rainfalls or temperature surges destroy the field surface. An additional impact of climate change on

soil erosion is limited vegetation growth that lowers field cover and exposes it to rains and winds.

- **Rainfall and Flooding:** High quantities of rain wash away topsoil particles, while big raindrops hit the field surface and destroy it with heavy splashes. Running currents during floods is another cause of soil erosion.
- **Wildfires:** Trees and shrubs reduce the speed of water run-offs. When forests or buffer zones are destroyed by wildfires, water streams have no obstacles in their way.

9.2. SOIL EROSION BY HUMAN ACTIVITIES

Apart from natural factors, soil erosion happens because of irresponsible farm management or deforestation to make way for urban area expansion, tourism development, road construction, and more.

9.3. FARMING PRACTICES THAT CAUSE SOIL EROSION

Natural vegetation is potentially the optimal protection over crops since farmlands are more vulnerable to rainfall and winds. Additionally, agricultural practices can lead to soil erosion because of a decrease of biodiversity both in vegetation variety and soil microorganisms. A lack of organic matter and beneficial biota has a negative impact on field fertility since not only earth particles but nutrients are removed from bare fields.

CHAPTER 10

EFFECTS OF SOIL EROSION ON NUTRIENT LEVELS IN FOOD CROPS

Contents

Soil erosion is a major threat to the nutrient levels in food crops, and it can have a significant impact on the overall quality of the food we consume. When soil erosion occurs, valuable nutrients are washed away or depleted from the soil, leading to lower nutrient levels in the crops that are grown in that soil.

10.1 NUTRIENT AVAILABILITY

One of the key ways that soil erosion affects nutrient levels in food crops is by reducing the availability of essential nutrients like nitrogen, phosphorus, and potassium. These nutrients are crucial for plant growth and development, and when they are lost through erosion, the crops that are grown in that soil may not receive the nutrients they need to thrive. This can result in lower nutrient levels in the food that is produced, which can have negative implications for human health.

In addition to reducing the availability of essential nutrients, soil erosion can also disrupt the balance of other important nutrients in the soil, such as calcium, magnesium, and zinc. These nutrients play a critical role in plant health and can impact the overall nutrient content of food crops. When soil erosion occurs, these nutrients can be lost or displaced, leading to imbalances that can affect the quality of the food that is grown in that soil.

Overall, soil erosion can have a significant impact on nutrient levels in food crops, and it is important for food scientists, university students, researchers, and policy makers to understand the effects of erosion on soil quality and nutrient content. By implementing sustainable agriculture practices and using soil amendments like compost, we can help to improve soil health and maintain nutrient levels in food crops for a healthier food supply.

Soil erosion lowers the agricultural value of lands through physico-chemical degradations. Soil nutrient loss due to runoff and sediment is a significant factor for soil fertility decline (Kurothe et al., 2014; Sahoo et al., 2015). The eroded sediments or soil are rich in crop nutrients, which are washed away from farmlands. Erosion-based constraints added to unfavorable climatic conditions will dictate the level of productivity of farming systems in sub-Saharan Africa (SSA). Soil erosion results in extreme losses of economic and environmental resources which have a negative impact on the economies of affected areas (Govers et al., 2016; Vaezi et al., 2017). On-site consequences are directly seen on yield in addition to soil properties, adversely affecting the ability of the soil to answer to management practices with time. The quantity of nutrients moved from croplands depends on the available soil handling practices and the selected farming system.

The nutrients moved via plant harvest (yield and crop leftovers) added to nutrient loss from erosion (runoff and sediment) are key threats to soil nutrient depletion in SSA and dictate the state of soils within the region. As an outcome, the majority of soils in SSA are strongly degraded such that specific integrated management practices involving organic and inorganic amendments are needed.

The nutrients lost via the soil erosion process can be quantified economically to reflect the effect of erosion on fertilizer investment. The loss of soil nutrients through erosion translates into significant cost because of the need for replacement to strengthen sustainability of cropping systems. In small-scale farming setups, this cost is ignored because of the lack of relevant information (Garcia-Doaz et al., 2017). Therefore, its quantification can help different stakeholders to adopt the most effective soil and crop management practices to minimize loss and improve crop output (Bertol et al., 2017).

CHAPTER 11

STRATEGIES FOR PREVENTING SOIL EROSION AND MAINTAINING NUTRIENT-RICH HARVESTS

Contents

Soil erosion is a major concern for farmers and gardeners alike, as it can lead to loss of valuable topsoil and nutrients essential for plant growth. In order to prevent soil erosion and maintain nutrient-rich harvests, there are several strategies that can be implemented.

One of the most effective ways to prevent soil erosion is through proper soil management practices, such as planting cover crops, contour plowing, and terracing. Cover crops help to protect the soil from erosion by holding it in place with their roots, while contour plowing and terracing help to slow down the flow of water and prevent it from washing away valuable topsoil.

Maintaining soil pH levels is also crucial for ensuring nutrient-rich harvests. Different plants require different pH levels in order to thrive, so it is important to test the soil regularly and adjust the pH as needed. This can be done through the addition of lime or sulfur to raise or lower the pH, respectively.

11.1 ORGANIC FARMING PRACTICES

Organic farming practices have been shown to improve soil quality and nutrient content in food crops compared to conventional farming methods. Organic farmers typically use natural fertilizers, such as compost and manure, to enrich the soil and promote plant growth. They also avoid the use of synthetic pesticides and herbicides, which can harm soil microbiology and impact nutrient availability for plants.

In addition to organic practices, the use of compost and other soil amendments can help to improve nutrient content in food crops. Compost is rich in organic matter and essential nutrients, making it an excellent way to replenish the soil and promote healthy plant growth. Other soil amendments, such as bone meal, kelp meal, and rock phosphate, can also be used to provide specific nutrients that may be lacking in the soil.

11.2 SUSTAINABLE AGRICULTURE

Sustainable agriculture practices are essential for maintaining soil quality and nutrient content in food crops. These practices focus on long-term environmental health and productivity, rather than short-term gains. This includes crop rotation, reduced tillage, and integrated pest management, which help to build soil health and promote nutrient-rich harvests.

Soil testing methods can be used to assess nutrient levels in food crops and determine if any deficiencies are present. This can help farmers and gardeners make informed decisions about soil amendments and fertilizer applications, in order to ensure that plants have access to the nutrients they need for optimal growth.

Overall, the relationship between soil health, plant health, and nutrient content in food products is clear. By implementing these strategies for preventing soil erosion and maintaining nutrient-rich harvests, farmers and gardeners can ensure that their crops are healthy and nutritious for consumption.

CHAPTER 12

SOIL HEALTH

Contents

The degradation of soil has an impact on human nutrition and health via its adverse impacts on the amount and quality of the food produced. A reduction in crops' yields and agronomic production increase food-insecurity that currently affects 854 million people worldwide. Minimal concentration of protein and micronutrients (including Zn, Fe, Se, B, I) enhance malnutrition and hidden hunger that affects 3.7 billion people, including children. Soil degradation lowers crop yields by augmenting susceptibility to drought stress and elemental imbalance. Strategies to address this situation cover:

- improving water productivity
- enhancing soil fertility and micronutrient availability
- adopting no-till farming and conservation agriculture

There are additionally new innovations including the application of remote sensing of plant nutritional stresses for targeted interventions, applying zeolites and nanoenhanced fertilizers and delivery systems, enhancing biological nitrogen fixation and mycorrhizal inoculation, conserving and recycling water using drip/sub-drip irrigation etc. Judiciously managed and properly restored, world soils have the capacity to support adequate and nutritious food for present and future communities.

Soil health refers to the continued ability of soil to function as an essential living ecosystem that maintains plants, animals, and humans. Healthy soil provides clean air and water, rich crops and forests, productive grazing lands, a diversity of wildlife, and beautiful landscapes. Soil does all this by carrying out five essential functions:

- **Managing Water:** Soil aids with controlling where rain, snowmelt, and irrigation water is directed to. Water flows over the land or into and through the soil.
- **Sustaining Plant and Animal Life:** The diversity and productivity of living organisms depends on soil.
- **Filtering and Buffering Potential Pollutants:** The minerals and microbes found as part of soil are responsible for filtering, buffering, degrading, immobilizing, and detoxifying organic and inorganic materials, such as industrial and municipal by-products and atmospheric deposits.
- **Nutrient Cycling:** Carbon, nitrogen, phosphorus, and many other nutrients are stored, transformed, and cycled in the soil.
- **Providing Physical Stability and Support:** Soil structure gives a medium for plant roots. Soils also give support for human structures and protection for archeological treasures.

Soil health research has established the way to manage soil in a way that enhances soil function.

- Maximize presence of living roots
- Minimize disturbance
- Maximize soil cover
- Maximize biodiversity

As the global population and demand for production of food increase, maintaining soil health and productivity is of paramount importance. By using soil health principles and systems that cover no-till, cover cropping, and diverse rotations, ever more farmers are improving their soil's organic matter and enhancing microbial activity. As an outcome, these practices allow for greater sequestering of carbon, increasing water infiltration, improving wildlife and pollinator habitat—all while harvesting better profits and often better crops.

Soil health should not only be determined by measuring crop yield, water quality, or any other single outcome. Soil health should not be measured directly; thus it is best to evaluate indicators. Indicators are quantifiable properties of soil or plants that give hints regarding how well the soil is able to perform. Indicators may be physical, chemical, and biological characteristics, processes, or properties of soils. They can also be shapes or visual displays of plants.

Useful indicators:

- are simple to measure
- quantify changes in the functions of soil
- cover chemical, biological, and physical properties
- are available to numerous users and applicable to field conditions
- are sensitive to changes in climate and management

Indicators can be evaluated by qualitative or quantitative techniques. Following the collection of measurements, indicators can be evaluated by looking for patterns and comparing results to measurements taken at a different time or field.

12.1. INDICATORS AND CONNECTION TO SOIL HEALTH

- **Physical:** bulk density, infiltration, soil structure and macropores, soil depth, and water holding capacity. This leads to the retention and mobility of water and nutrients; habitat for soil microbes; estimate of crop productivity potential; compaction, plow pan, water movement; porosity; and tilth.

- **Chemical:** electrical conductivity, reactive carbon, soil nitrate, soil pH, and extractable phosphorus and potassium. This leads to biological and chemical activity thresholds; plant and microbial activity thresholds; and plant available nutrients and potential for N and P loss.
- **Biological:** earthworms, microbial biomass C and N, particulate organic matter, potentially mineralizable N, soil enzymes, soil respiration, and total organic carbon. This leads to microbial catalytic potential and repository for C and N; soil productivity and N supplying potential; and microbial activity measure.

Soil works best when using management practices that increase soil health and enhance productivity and profitability immediately and into the future. An optimally functioning soil produces the greatest amount of outputs at the lowest cost. Maximizing soil health is key to maximizing profitability.

Managing improved soil function is in large part a matter of keeping suitable habitat for the diversity of creatures that makeup the soil food web. This can be achieved by disrupting the soil as little as possible, having as many different species of plants as practical, maintaining living plants in the soil as often as possible, and keeping the soil covered at all times.

12.2. Manage More by Disturbing Soil Less

Soil disturbance may be the outcome of physical, chemical or biological activities. Physical soil disturbance, including tillage, results in bare and/or compacted soil that is destructive to microbes living in the soil, and it develops a hostile environment for them to live. Misapplication of farm inputs may alter the symbiotic relationships between fungi, other microorganisms, and plant roots. Overgrazing, a type of biological disruption, decreases root mass, augments runoff, and increases soil temperature. All forms of soil alterations reduce habitat for soil microbes and produce a diminished soil food system.

12.3. Diversify Soil Biota with Plant Diversity

Plants utilize sunlight to transform carbon dioxide and water into carbohydrates that function as the building blocks for roots, stems, leaves, and seeds. They also function in combination with set soil microbes by liberating carbohydrates via their roots into the soil to act as food for the microbes in exchange for nutrients and water. A variety of plant carbohydrates is needed to support the collection of microorganisms in the soil. So as to reach an elevated level of diversity, various plants must be grown. The key to achieving soil health is guaranteeing that food and energy chains and webs are made up of various types of plants or animals, not simply one or two. Biodiversity is the key to the success of any agricultural system.

12.4. Keep a Living Root Growing Throughout the Year

Living plants keep a rhizosphere, an area of concentrated microbial activity in proximity to the root. The rhizosphere is the most active part of the soil ecosystem since it is where the majority of the readily available food can be found, and where peak nutrient and water cycling take place. Microbial food is discharged by plant roots to attract and feed microbes that give nutrients to the plant at the root-soil connection where the plants can absorb them. Because roots that are alive represent the easiest source of food for soil microbes, growing long-season crops or a cover crop after a short-season crop, feeds the foundation species of the soil food system as much as possible during the growing season.

Healthy soil is reliant on how well the soil food system is fed. Ensuring sufficient easily accessible food for soil microbes aids them in nutrient cycling that plants require to grow. Sugars from living plant roots, recently dead plant roots, crop residues, and soil organic matter all provide a food source for the many and varied members of the soil food system.

12.5. Keep the Soil Covered as Much as Possible

Soil cover preserves moisture, lowers temperature, catches raindrops to lower their destructive effect, limits weed growth, and acts as a habitat for members of the soil food system that spend at minimum some of their time above ground. This is the case independent of land use. Keeping the soil covered while permitting crop residues to break down, so their nutrients can be cycled back into the soil, is a complex process. Farmers must pay careful attention to their crop rotation (including any cover crops) and residue management if they are to maintain the soil covered and nourished at the same time.

CHAPTER 13

THE RELATIONSHIP BETWEEN SOIL PH AND NUTRIENT AVAILABILITY

Contents

Soil pH plays a crucial role in determining the availability of nutrients to plants, ultimately impacting the nutrient content of the food we consume. The pH level of soil directly affects the solubility and mobility of essential nutrients such as nitrogen, phosphorus, and potassium.

In acidic soils with a pH below 6.0, certain nutrients like phosphorus, calcium, and magnesium become less available to plants. On the other hand, alkaline soils with a pH above 7.5 can lead to deficiencies in micronutrients such as iron, zinc, and manganese. Maintaining a balanced pH level in the soil is essential for optimal nutrient uptake by plants and, consequently, nutrient-rich food production (Marschner, 2012).

Organic farming practices tend to have a more neutral pH level compared to conventional methods, which often rely on synthetic fertilizers that can alter soil pH over time. Additionally, sustainable agriculture practices, including the use of compost and other soil amendments, can help improve soil structure and pH balance, leading to enhanced nutrient availability for plants.

Soil erosion is another factor that can impact nutrient content in crops by depleting essential minerals from the soil. Implementing erosion control measures and promoting soil conservation practices are essential for maintaining soil quality and nutrient content in food production.

13.1 SOIL MICROBIOLOGY

Furthermore, the role of soil microbiology in nutrient availability cannot be overlooked. Beneficial microorganisms help break down organic matter, releasing nutrients for plants to absorb. Understanding and promoting a healthy soil microbiome is key to maximizing nutrient uptake in crops.

The relationship between soil pH and nutrient availability is a critical aspect of ensuring nutrient-rich harvests. By implementing sustainable agriculture practices, improving soil health, and monitoring nutrient levels, we can enhance the nutrient content of our food and promote overall human health and well-being.

Optimal soil pH range for best plant growth differs from crop to crop. Overall, soil pH 6.0–7.5 is acceptable for the majority of plants as the majority of nutrients become available in this pH window. Soil pH can be measured by mixing a sample of the soil with water and then evaluating the resulting solution.

One of the most important soil properties is its pH level since it influences the availability of nutrients, thus having an impact on plant growth. Keeping soil pH in the desired level is key for success as plant growth is deficient when the soil pH is outside of the wanted range. A huge number of chemical, biochemical, and biological reactions take place in the soils. By way of these reactions, nutrients, whether already present in the soil or added by fertilizers, are transformed into forms that can be absorbed by plant roots. The pH of the soil has an impact on these reactions and in doing so, dictates nutrient availability essential for plant growth. This is done by affecting a nutrient's solubility in the water that is present in the soil. The more soluble a nutrient is, the more that is dissolved in the soil water. Plants absorb water from the soil with the dissolved nutrients in it via their roots. This is how plants obtain most of the nutrients required for growth and development. Soil pH also has an impact on the toxicity of certain elements and influences soil microbial activity. For example, the metabolization of organic matter is lower in acidic soil because growth and reproduction of the soil microbes, mainly bacteria and fungi, are limited.

Many soils have a pH in the range of 4.5 to 5.5 while many plants, including vegetables, flowers and turf grasses, prefer the soil pH to be between 6.0 and 6.8. Some notable exceptions include blueberries and broad-leaved evergreens such as rhododendrons, which need acid soils for good growth. In fact, it is not the acidity that these ericaceous plants require, rather it is iron. These plants have greater iron requirements than many other types of plants and iron is soluble in the soil water when the pH is low and, thus, more available to the plants.

When the soil pH goes below 6.0, nutrients such as phosphorus, nitrogen and potassium turn less available for plants. Acid soils are also usually deficient in magnesium and calcium, two key plant nutrients. Another challenge with acid soils is that elements such as aluminum are much more soluble and may be absorbed in amounts that can damage plants. One plant that is able to tolerate aluminum well is the blue hydrangea. Aluminum is required for that characteristic blue color.

On the other hand, a soil pH that is higher than 7.5 can also turn nutrients unavailable. While there are regions that have naturally high pH calcareous soils, this is not usually the case. Usually, an elevated soil pH in vegetable or flower beds is an indicator of heavy additions of limestone, wood ashes or compost.

13.2 SOIL ACIDIFICATION

The acidification of soil can happen because of organic matter decomposition (e.g., crop residues, manure), the use of fertilizers with ammonium (including ammonium sulfate, monoammonium phosphate, diammonium phosphate), and the leaching of base cations (such as magnesium or calcium) from the topsoil in the case of rain or via irrigation, leaving acidic cations, particularly aluminum, as the main exchangeable cation. Limestone is usually the material of choice to increase the pH of soils. It neutralizes soil acidity while also providing required calcium and often magnesium. Dolomitic limestone, which has both of these elements, is most widely available in some North American locations and most often recommended. If the magnesium level of the soil is above desired, a calcitic limestone, which is composed mostly of calcium compounds, could help.

Hydrated lime is fast-acting however quite caustic and is not recommended for use outside commercial entities. Wood ashes may also be employed as a liming agent at approximately 1 ½ times the rate of the recommended limestone application. Leaving wood ashes on plant leaves may injure them. It is recommended to lightly water wood ashes into the soil or add them prior to planting.

13.3 CHANGING SOIL PH

There are numerous products in the market that claim to be fast acting and change the pH in short order. Research has not found these products to elicit a pH change much faster than ground or pelletized limestone. In addition, package instructions suggest that these materials must be applied at regular intervals to maintain the soil pH level.

Sometimes, lowering a soil's pH to accommodate the growing requirements of specific plants is required. Usually, this is required when plants that evolved to tolerate low pH soils, including rhododendrons, azaleas, blueberries, and oaks are planted in soils with a pHabove 5.0. Sulfur or aluminum sulfate are two of the most common materials employed to reduce soil pH.

Soil testing is the best approach to determine how much limestone and/or fertilizer to use. Late summer and until fall is the optimal time to test the soil since any limestone or sulfur recommended and applied will have time to act on the soil pH prior to the next planting season. Nevertheless, soils can be tested and limestone/sulfur applied any time the ground is not frozen.

CHAPTER 14

EFFECTS OF SOIL PH ON PLANT UPTAKE OF NUTRIENTS

Contents

Soil pH plays a crucial role in determining the availability of nutrients for plant uptake. Different nutrients have varying levels of availability at different pH levels, with some becoming more or less accessible to plants depending on the acidity or alkalinity of the soil.

In acidic soils, essential nutrients such as phosphorus, calcium, and magnesium may become less available to plants due to the high levels of hydrogen ions present. This can lead to nutrient deficiencies in crops, impacting their growth and overall nutrient content. On the other hand, in alkaline soils, micronutrients like iron, manganese, and zinc may become less available, also affecting plant growth and nutrient uptake.

Maintaining a balanced soil pH is essential for ensuring optimal nutrient availability for plants. Organic farming practices often focus on maintaining neutral pH levels through the use of natural amendments like compost, manure, and cover crops. These practices not only help to improve soil structure and microbial activity but also enhance nutrient availability for plants.

In comparison, conventional farming practices may rely more heavily on synthetic fertilizers, which can alter soil pH levels over time. This can lead to imbalances in nutrient availability and potential issues with nutrient uptake in crops.

By understanding the effects of soil pH on plant nutrient uptake, farmers, researchers, and policymakers can work together to implement sustainable agriculture practices that prioritize soil health and nutrient content in food. Soil testing methods can be used to assess nutrient levels in crops and guide decision-making on soil amendments and fertilizers to maintain optimal soil pH levels and nutrient availability for healthy plant growth and nutrient-rich harvests.

14.1 MANAGING SOIL PH LEVELS

Managing soil pH is crucial for maintaining optimal nutrient content in food crops. Soil pH levels can greatly impact the availability of essential nutrients for plants, ultimately affecting the nutrient content of the food we consume. Here are some strategies for managing soil pH to ensure nutrient-rich harvests:

- **Regularly Test Soil pH:** It is important to regularly test the pH levels of your soil to determine if any adjustments are needed. Soil testing kits are readily available and can help you monitor and maintain the pH of your soil.
- **Adjust Soil pH as Needed:** Depending on the results of your soil tests, you may need to adjust the pH of your soil. Adding lime can

help raise pH levels in acidic soils, while sulfur can lower pH levels in alkaline soils.

- **Use Organic Farming Practices:** Organic farming practices can help maintain soil pH levels naturally. Avoiding synthetic fertilizers and chemicals can help preserve the natural balance of soil pH.
- **Incorporate Compost and Soil Amendments:** Adding compost and other organic soil amendments can help improve soil structure and pH levels. Compost is rich in nutrients and can help maintain a healthy pH balance in the soil.
- **Practice Sustainable Agriculture:** Sustainable agriculture practices, such as crop rotation and cover cropping, can help maintain soil pH levels and promote overall soil health. These practices can also help reduce erosion and preserve nutrient content in food crops.

By implementing these strategies for managing soil pH, food scientists, university students, researchers, and policy makers can work towards ensuring nutrient-rich harvests and promoting the link between soil quality and food nutrient content.

CHAPTER 15

EVALUATION OF SOIL QUALITY

Contents

15.1. BENEFITS OF COMPOST AND SOIL AMENDMENTS FOR ENHANCING SOIL QUALITY

Composting is a controlled, aerobic (oxygen-required) process that converts organic materials into a nutrient-rich, biologically-stable soil amendment or mulch through natural decomposition. The end product is compost. Microorganisms feed on the materials added to the compost pile during the composting process. They use carbon and nitrogen to grow and reproduce, water to digest materials, and oxygen to breathe.

You can compost at home using food scraps from your kitchen and dry leaves and woody material from your yard.

Compost and soil amendments play a crucial role in enhancing soil quality, which ultimately leads to nutrient-rich harvests. By adding organic matter to the soil through composting, farmers can improve soil structure, increase water retention, and promote beneficial microbial activity. This results in healthier plants with access to a wider range of essential nutrients, leading to higher nutrient content in the food they produce.

One of the key benefits of using compost and soil amendments is their ability to balance soil pH levels. Soil pH plays a significant role in nutrient availability to plants, with certain nutrients becoming more or less available at different pH levels. By maintaining optimal pH levels through the use of compost and soil amendments, farmers can ensure that plants have access to the nutrients they need for healthy growth and development.

Organic farming practices, which often include the use of compost and other soil amendments, have been shown to have a positive impact on soil quality. Compared to conventional farming practices, organic farming promotes soil health by reducing erosion, improving soil structure, and fostering a diverse microbial community. This, in turn, leads to higher nutrient content in the food produced on organic farms.

Overall, the use of compost and soil amendments is essential for sustainable agriculture practices that maintain soil quality and nutrient content. By understanding the relationship between soil health, plant health, and nutrient content in food products, farmers and researchers can work together to improve the quality of our food supply for the benefit of all.

Composting is nature's way of recycling. It is one of the most powerful actions we can take to reduce our trash, address climate change, and build healthy soil. By turning our food scraps and yard trim into compost, we transform our waste streams into a beneficial, value-added soil amendment and use it to protect the environment and create resilient communities.

- Composting is a resourceful way to recycle the food scraps and yard trim you generate at home all year and manage your waste more sustainably.
- You reduce the volume of materials that might otherwise be disposed of in landfills or trash incinerators – leaves, grass clippings, yard trim, and food scraps – and prevent powerful greenhouse gases from being emitted into the atmosphere.
- Composting involves minimal effort, equipment, expense, and expertise and can be fun.
- You save money by producing a free, high-quality soil amendment – compost – which reduces your use of fertilizer and pesticides.
- You can use your compost to build healthier soil, prevent soil erosion, conserve water, and improve plant growth in your garden and yard.

15.1.1. Steps for Backyard Composting

1. Determine how you will collect and store your browns and greens. Collect and store your fruit and vegetable scraps in a closed container on your kitchen counter, under your sink, or in your fridge or freezer. For browns, set aside an area outside to store your steady supply of leaves, twigs, or other carbon-rich material (to mix with your food scraps).
2. Set aside space for your compost pile and build or buy a bin. Choose a space in your yard for your compost pile that is easily accessible year-round and has good drainage. Avoid placing it right up against a fence and ensure there is a water source nearby. Your compost pile will break down in sun or shade. Next, choose a type of bin for your pile. Bins can be constructed from materials such as wire, wood, and cinder blocks. They can also be enclosed and include barrels and tumblers.
3. Prepare your ingredients for composting. Before adding your browns and greens to the pile, try to chop and break them up into smaller pieces (e.g., corn cobs, broccoli stalks, and other tough food scraps). Doing so will help the materials in the pile break down faster.
4. How to build your compost pile. Start your pile with a four- to six-inch layer of bulky browns such as twigs and wood chips. This layer absorbs extra liquids, elevates your pile and allows air to circulate at the base of the pile. Then layer your greens and browns like lasagna. If needed, add a little water to dampen the pile. Having the right proportions of ingredients in your compost pile will provide

the composting microorganisms the carbon, nitrogen, oxygen, and moisture they need to break down the materials into finished compost. When adding browns and greens to your pile, add at least two to three times the volume of browns (such as dry leaves) to the volume of greens (such as food scraps). Always ensure your food scraps are covered by four to eight inches of dry leaves or other browns. Air and water are the other key ingredients in your pile. To ensure air circulation, add enough browns and turn your compost occasionally. To maintain moisture in your pile, ensure your combined materials have the consistency of a wrung-out sponge.

5. Maintain your compost pile. As the materials in your compost pile begin to decompose, the temperature of the pile will initially begin to rise, especially in the center. A backyard pile, if well maintained, can reach temperatures of 130° to 160° F. High temperatures help reduce the presence of pathogens and weed seeds. Turning and mixing your pile from time to time will help speed up the decomposition process and aerate the pile. Use a garden fork to turn the outside of the pile inward. Monitor your pile for moisture, odor, and temperature and make adjustments as needed.
    a. If the pile is too dry, activity in the pile will slow or cease. Moisten the pile and turn it (Refer to the note above about maintaining moisture in your pile)
    b. If the pile has a bad odor, it may be too wet or need more air circulation. Add more browns/dry material to the pile and turn the pile.
    c. If the pile is not heating up, mix in greens and turn the pile.
6. Harvest your finished compost. When your compost pile is no longer heating up after mixing, and when there are no visible food scraps, allow your pile to cure, or finish, for at least four weeks. You can relocate the oldest compost at the bottom of the pile to a separate area to cure or stop adding materials to your pile. After curing, your pile will shrink to about one-third of its original size. Compost in a well-maintained pile will be finished and ready for use in about three to five months. Left untended, a pile may take a year to decompose. The compost will look dark, loose, and crumbly and smell like fresh soil. Most, if not all, of the materials that went into the compost pile should be decomposed. Screen or sift your finished compost to filter out materials that didn't break down – twigs, fruit pits, eggshells, and items like produce stickers and plastic (You can make a homemade screener out of ¼ inch hardware cloth). Pits, eggshells, etc., that you sifted out can be added back into the active pile or to a new pile.

15.1.2. Avoiding Rodents

Home compost piles that are well constructed and maintained properly should not attract pests or rodents.

- If using a bin, it is important to reinforce the bin with a lid and potentially a floor. There should be no holes or gaps in your bin that are larger than ¼ inch.
- Maintain the proper ratio of materials in your pile: two to three parts carbon-rich material (browns) to one part nitrogen-rich material (greens).
- Ensure you cover and bury your food scraps in your pile.
- Do not add meat, dairy, or greasy foods to your pile.

15.1.3. Worm Composting (Vermicomposting)

Worm composting, or vermicomposting, is another method of composting you can try at home. It takes up little space, the materials are simple and inexpensive, and it can be done indoors or outdoors. You will need a container or bin, bedding material, worms, and food scraps. If your worm bin is properly maintained, it should not emit odors or attract pests. The resulting product is vermicompost, a soil amendment.

Steps for Vermicomposting

1. **Make or Buy a Worm Bin:** You can purchase a worm bin or make your own of untreated wood or plastic storage bins. The bin should have a tight-fitting lid and be a dark color to keep out light. If using plastic storage containers, drill air holes around the upper sides of one bin near the lid and drainage holes on the bottom of the same bin. Place the bin with the drilled holes inside the other bin (The bottom bin will catch any liquid that drains out of the top bin).
2. **Choose a Space for Your Worm Bin:** You can keep your worm bin indoors (e.g., under a sink, in a closet or basement) or outdoors in the shade. If your bin is outdoors (e.g., garage, carport, porch, deck, or apartment balcony), insulate the bin with blankets, straw, or other materials to keep it warm during colder months. You want to maintain a bin temperature of 59° to 77°F; however, the worms can survive at 32° to 95° F if they have at least four inches of bedding.
3. Materials for Your Bin:
    a. **Worms:** Of the 9,000 species of earthworms, only seven are suitable for vermicomposting. One of the most used is the "red wriggler" or *Eisenia fetida*. It is essential to use a suitable species. Do not use worms you might find outdoors

or purchase at a bait shop. Begin your bin with one pound of worms, about 1,000 individual worms. Purchase them from a worm grower or find a neighbor in your community who has a vermicomposting bin and can give you some.

b. **Bedding:** Shred or tear non-glossy newspaper, office paper, cardboard, or dry leaves and soak the bedding material for 10 minutes. Wring it out so it feels like a moist sponge. Place it in your bin and fluff it up. Fill the bin almost halfway with the moist bedding and then add a handful of soil.

c. **Food:** You can feed your worms most fruit and vegetables scraps, coffee grounds and paper filters, crushed eggshells, paper tea bags without staples, and crushed eggshells. Avoid citrus fruits, odorous foods like onions and garlic, meat, dairy products, greasy foods, bones, and pet waste. Chop up scraps into small pieces before adding them to the bin to allow for faster decomposition.

4. **Start Your Bin and Feed Your Worms:** Place worms on top of the bedding in the bin. Once the worms have settled in the bin, add some food scraps on the surface of the bedding. Each time you add food scraps, make sure to cover them with two inches of bedding. Adding the proper ratio of materials to the bin and covering them is important. Worms eat about 25% of their weight each day. As worms reproduce quickly, you can increase the amount you feed them. Always ensure the food scraps have been eaten before adding more.

5. **Harvest and Use Your Vermicompost:** After three to six months, you will be able to harvest your vermicompost, or worm castings, at the bottom of your worm bin. Vermicompost doesn't need to cure and can be used immediately or stored for future use.

You can add compost to your flower and vegetable beds, window boxes, and container gardens; incorporate it into tree beds; mix it with potting soil for indoor plants; or spread it on top of the soil in your yard.

Compost can be used as a soil amendment or as a mulch. As a soil amendment, mix in two to four inches of compost to the top six to nine inches of your soil. As a mulch, loosen the top two to three inches of soil and add a three-inch layer of compost on the surface, a few inches away from plant stems and tree trunks.

Adding finished compost to your soil:

- Improves the structure and health of your soil by adding organic matter.

- Helps the soil retain moisture and nutrients.
- Attracts beneficial organisms to the soil and reduces the need for pesticides and fertilizers.
- Reduces the potential for soil erosion.
- Sequesters carbon in the soil.
- Builds resiliency to the impacts of climate change.

15.2. TECHNIQUES FOR INCORPORATING COMPOST INTO FARMING PRACTICES

Incorporating compost into farming practices is a crucial technique for improving soil quality and ultimately increasing the nutrient content of food crops. Compost is a valuable source of organic matter, which helps to improve soil structure, retain moisture, and enhance nutrient availability for plants. For food scientists, university students, researchers, and policy makers interested in maximizing the nutrient content of food, understanding how to effectively incorporate compost into farming practices is essential. One technique for incorporating compost into farming practices is to use compost as a soil amendment before planting crops. By mixing compost into the top layer of soil, farmers can improve soil fertility and provide plants with a rich source of nutrients. Additionally, compost can be used as a mulch around plants to help retain moisture and suppress weeds, further enhancing plant growth and nutrient uptake.

Another technique is to create compost tea, a liquid fertilizer made by steeping compost in water. Compost tea can be applied directly to plant roots or sprayed on leaves to provide a quick boost of nutrients and beneficial microorganisms. This can help improve plant health and increase nutrient uptake, resulting in higher nutrient content in food crops.

Overall, incorporating compost into farming practices is a sustainable and effective way to improve soil quality and increase the nutrient content of food crops. By utilizing compost as a soil amendment or creating compost tea, farmers can enhance plant growth, improve nutrient availability, and ultimately produce nutrient-rich harvests. For those interested in soil quality, nutrient content of food, and sustainable agriculture practices, incorporating compost is a valuable technique to consider.

15.3. COMMON SOIL TESTING METHODS AND THEIR APPLICATIONS

The quality of soil plays a key function in establishing the nutrient content of food crops. Numerous soil testing methods are utilized to measure the nutrient levels in the soil and ultimately improve the quality of the food to be eaten. In this section, some commonly used soil testing methods and their applications in understanding the link between soil quality and food nutrient content are discussed.

One of the most regularly used soil testing methods is pH levels. Soil pH levels have a strong impact on the nutrient amount of food crops. Different plants thrive at different pH levels, thus the importance of adjusting the soil pH respectively to ensure best nutrient absorption by the plants.

Soil microbiology analysis is another key soil testing method. Soil microbiology plays a key role in nutrient availability for plants. By evaluating the microbial population in the soil, scientists can establish the nutrient cycling processes and take informed decisions on soil amendments to optimize nutrient content in food crops.

The type of farming practices (organic vs conventional) have an additional significant impact on the quality of the soil. Organic farming practices target the building of soil health via the utilization of compost and other organic amendments. These practices help maintain soil structure and fertility, resulting in nutrient-rich yields.

The erosion of soil is another characteristic that affects nutrient content in crops. Eroded soils are usually depleted of essential nutrients, resulting in lower nutrient content in food plants. Sustainable agriculture practices, including cover cropping and conservation tillage, can aid in the prevention of soil erosion and keep soil quality for nutrient-rich harvests.

Soil testing methods are basic tools for measuring nutrient levels in food crops and grasping the relationship across soil health, plant health, and nutrient content in foods. By setting up sustainable agriculture practices and using soil amendments, it is possible to increase soil quality and achieve nutrient-rich harvests for a healthier food supply.

15.4. INTERPRETING SOIL TEST RESULTS FOR IMPROVING NUTRIENT CONTENT IN FOOD

Soil testing is a crucial tool for farmers, gardeners, and agricultural specialists to understand the nutrient levels in the soil and make informed decisions about how to improve them. By interpreting soil test results, it is possible to take steps to ensure that food crops are receiving the proper nutrients they need to flourish and provide nutrient-rich harvests.

One key factor to consider when interpreting soil test results is the pH level of the soil. Soil pH plays a significant role in determining the availability of essential nutrients to plants. For example, if the soil is too acidic or too alkaline, certain nutrients may become less available to plants, leading to nutrient deficiencies in the crops grown in that soil. By adjusting the pH levels through various methods such as adding lime or sulfur, it is possible to optimize nutrient availability for plant uptake.

Organic farming practices also play a significant role in improving soil quality and nutrient content in food. Compared to conventional farming methods, organic practices focus on building healthy soil ecosystems through the use of compost, cover crops, and crop rotations. These practices not only improve soil structure and fertility but also enhance the nutrient content of the food crops grown in such environments.

Soil erosion is another critical factor that can impact the nutrient content in crops. Erosion can lead to the loss of topsoil, which is rich in essential nutrients necessary for plant growth. By implementing soil conservation practices such as contour plowing, terracing, and planting windbreaks, it is possible to reduce soil erosion and maintain nutrient-rich soils for healthy crop production.

In addition to soil quality and erosion, soil microbiology also plays a vital role in nutrient availability for plants. Soil microbes such as bacteria and fungi help break down organic matter and release nutrients that plants can absorb. By fostering a healthy soil microbiome through practices like adding compost and other organic amendments, it is possible to enhance nutrient availability for food crops.

Overall, interpreting soil test results is essential for understanding the nutrient levels in our soils and improving nutrient content in food. By implementing sustainable agriculture practices, utilizing compost and soil amendments, and focusing on soil health, it is possible to ensure that crops are nutrient-rich and contribute to a healthy diet for consumers.

Box 1: Acid vs. Alkaline Soils

Soil pH is a key factor in farmland as it controls availability of nutrients, microbial activity and crop productivity. Before delving into what causes soils to become acid or alkaline and the steps to take to treat and correct soil pH, we must first establish what is considered an optimal pH for crop production.

The only sure way to determine the pH content of soil is to do a soil test. Soil pH measures hydrogen (H) ions in soil and typically ranges between 5.5 and 8.0. For most prairie crops, the ideal soil pH range for optimal growth and development is of 6.5 to 6.8.

Soils with pH ranging from 5.6 to 6.0 are considered moderately acid, while strongly acid and very strongly acidic soils have pH ranging from 5.1–5.5 to <5.0, respectively. Crops have difficulty establishing and show a decline in productivity and yield in soils with a pH below 6.0.

When soil pH is less than 5.5, the soil is very acidic and likely has high levels of aluminum and/or iron, alongside lower content of calcium and magnesium. Low pH reduces availability of nutrients such as phosphorus and molybdenum, affects nitrogen fixation and causes crop toxicity with elements such as aluminum or manganese that become more soluble at a lower pH. Soil liming may be necessary when the pH is less than 6.0. Additionally, root growth can increase by 40% when pH rises from 5.5 to 7.2.

15.5. ACIDIC SOILS

15.5.1. What Causes Soils to Become Acid?

Acidic soils are often high in iron and aluminum oxides, since they are the slowest minerals to weather in soil. Aluminum in these ever more acidic soils is solubilized and will mix with water to release additional hydrogen ions adding to further acidity. Soils in areas with high amounts of rainfall are often acidic since the water leaches basic cations including calcium, magnesium, potassium and sodium out of the soil profile, leaving room for acidic cations including hydrogen and aluminum.

The parent substrate from which the soil developed may be a source of acidity. For example, soils developed from land with high organic matter have conifers or with elevated amounts of iron or aluminum are likely to be acidic. Acidity can also be elevated by numerous other factors, such as the nitrification of ammonium fertilizers, which yields hydrogen ions.

15.5.2. Consequences of Soil Acidity

Low pH due to soil acidity augments the solubility of zinc, manganese, iron and aluminum, which can translate into rapid accumulation and toxicity. In addition, the elevated amounts of aluminum and iron in acid soil lead to a tie-up, lowering phosphorus availability; molybdenum turns less available to plants, which has an impact on nitrogen and phosphorus absorption and use, limits root growth and lowers nodulation in pulse crops; microbial activity, the breakdown of organic matter and mineralization are negatively impacted; and, in acid conditions, calcium, magnesium, and potassium deficiencies become heavier.

15.5.3. Improving Acidic Soils

An often used and lower cost manner to increase the pH of the soil is by liming. The quantity of lime material needed depends on the pH of the untreated soil and the targeted pH to produce the crop, the quantity of soluble and exchangeable acidity, the crop's tolerance to acidity/alkalinity, the quantity of organic matter in the soil, and the form of clay found in the soil. Lime is most effective at neutralizing acidity when it is added or tilled into the soil to the entire depth of the root area.

15.6. ALKALINE SOILS

Alkaline soil is the term utilized to describe a soil with an elevated pH level (higher than 7.3). When soil pH is higher than 8.0, access to nutrients such as phosphorus or micronutrients (i.e., Zn, Cu, Mn, Fe) can be lowered and soil pH higher than 8.3 can show high sodium or sodic soil challenges and often drainage problems. Elemental sulfur may be needed on these soils particularly if they are rich in magnesium.

15.6.1. What Causes Soils to Become Alkaline?

Soils become alkaline for a number of reasons. They might be found on very dry land and/or in areas with limited rainfall. The alkalinity might be due to the nature of the soil itself, or as an outcome of receiving water that has strongly alkaline substances (such as calcium or magnesium carbonate). Soil that has not ever been broken usually has a neutral pH yet once cultivated, it usually becomes alkaline due to calcium and other salts taken from the soil's lower horizons.

15.6.2. Consequences of Soil Alkalinity

Overall, alkaline soil has a rich calcium carbonate amount, which stops plants from taking up most of the other key nutrients. Regardless of adequate nutrients being present in the soil, plants are unable to use them. For example, soybeans grown on very alkaline soils tend to develop iron chlorosis, which can be easily corrected through the use of iron, supplied in-furrow or by foliar application.

15.6.3. Improving Alkaline Soils

so as to manage alkaline soil, farmers may try the following methods to reduce soil pH: add organic matter to the soil – the acidic reaction set off by decomposition will aid with the reduction of pH levels and the process will support with drainage and aeration; utilizing humic acid regularly to promote biological activity involved in mineralization and breakdown of organic matter, which in turn can create organic acids; incorporate sulfur or sulfur-coated urea into the fertility program.

CHAPTER 16

BIOFORTIFICATION

Contents

Biofortification refers to the process by which the density of nutrients in foods is elevated via conventional plant breeding, and/or enhanced agrological processes and/or modern biotechnology while not giving up any characteristic that is desired by consumers (Nestel, 2006). Biofortification is seen as a nutrition-sensitive-agriculture intervention that may lower vitamin and mineral deficiencies (Ruel and Alderman, 2013). Iron-biofortification of beans, cowpea and pearl millet, zinc-biofortification of maize, rice, and wheat, and pro-vitamin A carotenoid-biofortification of cassava, maize, rice, and sweet potato are ongoing and at different phases of implementation (Bouis et al., 2011; Saltzman et al., 2013).

The biological process for biofortified crops to help with the improvement of nutritional status is straightforward: biofortified plants used for food are more nutrient-dense than non-biofortified forms. Thus, assuming matching micronutrient bioavailability (La Frano et al., 2014) and retention (De Moura and Milof, 2015) after cooking or processing and storage, people will intake (Muñoz et al., 2008) and absorb (Rosado et al., 2009) higher amounts of micronutrients from consuming biofortified crops than from an equal quantity of non-biofortified crops. In people with a diet restricted in these micronutrients, opting for biofortified staple crops may improve micronutrient consumption.

Outcomes are optimistic for iron-biofortified crops, because partially iron-biofortified rice has increased the iron stores of reproductive-age women in the Philippines (Haas et al., 2005), iron-biofortified pearl millet has improved the iron stores and alleviated iron deficiency in school-aged children in India (Finkelstein et al., 2015) and iron-biofortified beans have increased the iron stores in women in Rwanda (Haas et al., 2016). Evidence for pro-vitamin A rich crops is harder to interpret as provitamin A carotenoids are initially absorbed in the body and then turned into the active form of vitamin A in accordance to the body's requirement for the nutrient. There is ample evidence (both efficacy and effectiveness) that provitamin A biofortified orange sweet potato lowers vitamin A deficiency in children in Mozambique (Low et al., 2007; Hotz et al., 2012), Uganda (Hotz et al., 2012b) and in South Africa (van Jaarsveld et al., 2005), with an extra study in Bangladesh showing increased pro-vitamin A concentration but not vitamin A status (Jamil et al., 2012).

Research with yellow cassava biofortified with provitamin A in Kenya reported a higher vitamin A status and a larger increment in pro-vitamin A concentrations in school-aged children (Talsma et al., 2016). Although additional trials using biofortified crops with greater provitamin A densities in other populations are in process, evidence is mounting in favor of biofortified crops being able to improve the micronutrient status of different populations. There are however, two major behavioral challenges linked to biofortification: one for the farmer and one for the consumer. Farmers are looking to plant new varieties that are agronomically better to the currently available types they plant, such as cultivars that are more drought resistant, have greater yield or lower susceptibility to diseases (Pfeiffer et al., 2007). Plants with better micronutrient concentration with similar or lower agronomic performance will not be an option for farmers (Bouis et al., 2011; Saltzman et al., 2013). Thus, biofortified varieties need to be agronomically equal or ideally superior to the less nutrient dense market and traditional varieties.

Provitamin A carotenoids give color to foods; thus, biofortification with these carotenoids will modify the color of crops. Therefore, motivating customers to modify their purchasing and consumption habits from white- to orange-fleshed maize, cassava, or sweet potato may be difficult (Robyn and Winter-Nelson, 2008). Nevertheless, other properties such as taste are also impacted by improved pro-vitamin A levels and can produce an overall better rating (Talsma et al., 2013). Additionally, acceptability has been demonstrated to improve by giving information about the health benefits of the biofortified version to the consumer (Birol et al., 2015).

Plants that are biofortified are usually assumed to be transgenic or genetically modified crops. Although the push of biofortification activities via international projects (Bouis et al., 2011; Saltzman et al., 2013) targeted the production of nutritionally improved crops through traditional processes, more projects are using genetic modification to increase value to staple and non-staple plants. Thus, the education of the consumer and a set regulatory framework to evaluate the benefits and risks of genetically modified crops will be required to add to consumer confidence in biofortified crops developed via genetic modification.

To achieve its full potential, biofortification needs to be integrated as a main activity within a range of global institutions. This will necessitate three basic elements.

- **Supply:** Agricultural scientific groups, both public and private, agree in the recognition of high mineral and vitamin content as key crop breeding goals; variety release expert groups establish minimum levels of minerals and vitamins a requirement for approval for release.
- **Policy:** A group of national and international public officials agree on the significant impact of biofortification for improving and maintaining public health, in addition to the high economic return of investments in biofortification and the legitimacy conferred by international recognition (particularly by standards bodies).
- **Demand:** Both rural and urban consumers come to see the value of, and demand, high mineral and vitamin content in their staple foods.

16.1. GENETIC ENGINEERING

When enough natural genetic variation in nutrient content is not present in a particular food crop, transgenic approaches represent an alternative method in achieving biofortification. This is achieved by introducing novel genes obtained from different organisms that are new to the target plant species, or that turn on or upregulate existing genes in the target tissue. Even though a vast amount of this research is in the early stages of development, proof-of-concept has been set for biofortification with numerous nutrients in a diversity of food crops (Table 16.1).

Biofortified staple food crops resulting from the employment of transgenic methods that are most advanced in development is rice that contains β-carotene. Even though rice plants have the genes needed to produce β-carotene, their expression is turned off in the rice grain endosperm. The insertion of two genes, a plant phytoene synthase (PSY) and bacterial carotene desaturase (CRTI) resulted as sufficient to turn on the pathway. Named Golden Rice, the initial proof-of-concept milled rice grain had only a limited quantity of β-carotene ($<1.6\ \mu g\ g^{-1}$ dry weight). Nevertheless, a second generation Golden Rice using PSY obtained from maize instead of daffodil has a greater β-carotene content of up to >30 µg g−1dry weight prior to storage, which is of higher nutritional significance.

Table 16.1. Biofortified Food Crops Being Developed Using Genetic Engineering

Nutrient	**Food Crop**
Calcium	Carrot, lettuce
Iron	Maize, rice
Folate	Rice, tomato
Provitamin A (β-carotene)	Cassava, maize, rice, potato, tomato
Vitamin E	Maize, oil seed crops
Amino acids	Cassava, maize, rice, sorghum, wheat
Fatty acids	Oil seed crops

CHAPTER 17

UNDERSTANDING THE ROLE OF SOIL MICROORGANISMS IN NUTRIENT CYCLING

Contents

Nutrient cycling is understood to mean that nutrients move among different components of a cell, community, or ecosystem and can be cycled and reutilized by some of these components. It is key to measure the speed at which nutrients are taken up, moved among biota, and liberated for subsequent reassimilation. This cycling speed per unit time is a characteristic full of information of community or ecosystem operations, especially in how it changes in response to alteration of physical and biotic parameters. Nutrients, nevertheless, move within an aquatic ecosystem along a gradient from land to water. Nutrients are lost from sites of photosynthetic and secondary production, and such losses must be restocked by nutrient inputs for sustenance, growth, and net production to continue.

17.1 NUTRIENT CYCLING

Nutrient cycling takes place at many spatial and temporal scales. These processes have to be considered as relatively self-contained in communities in which a high amount of recycling of nutrients is taking place. Many micro communities with high rates of recycling are connected along and within fewer larger-scale communities with progressively longer and slower speeds of recycling. Since growth and reproduction are preeminent among organisms, speeds of nutrient cycling and recycling are important to achieve metabolic demands. These rates of nutrient cycling are controlled by a complex interplay of physical, chemical, and biotic factors. Environmental and community factors controlling nutrient processes in lentic waters can be alike to some of those for flowing waters, but differ mostly in greater rates and interactive intensities due to the much higher photosynthetic productivity, loading of organic matter, and habitat diversity among lentic waters than is the case in lotic waters (Wetzel, 1983b,c; Wetzel and Ward, 1992).

Analysis of the importance of nutrient cycling and recycling needs foremost an appreciation that the habitat conditions in lentic (non-riverine) fresh waters dictate nutrient conservation and retention in order to reach and sustain the very high levels of community productivity.

Soil microorganisms play a crucial role in the nutrient cycling process, which is essential for maintaining soil quality and ensuring the nutrient content of food crops. These tiny organisms, including bacteria, fungi, and protozoa, break down organic matter in the soil, releasing nutrients such as nitrogen, phosphorus, and potassium that are vital for plant growth.

One of the key functions of soil microorganisms is their ability to decompose organic matter, such as dead plant material and animal waste. During this decomposition process, microorganisms release enzymes that break down complex organic compounds into simpler forms that plants can absorb. This results in a continuous supply of nutrients for plant uptake, promoting healthy growth and development.

In addition to decomposing organic matter, soil microorganisms also play a role in fixing nitrogen from the atmosphere into a form that plants can use. Certain bacteria, known as nitrogen-fixing bacteria, have the ability to convert atmospheric nitrogen gas into ammonium, which can then be taken up by plants. This process is essential for nitrogen cycling in the soil and is crucial for the growth of nitrogen-demanding crops such as legumes.

Furthermore, soil microorganisms help to regulate nutrient availability in the soil by competing with plants for nutrients. By breaking down organic matter and releasing nutrients slowly over time, microorganisms help to prevent nutrient leaching and ensure a steady supply of nutrients for plants. This balance is essential for maintaining soil fertility and promoting healthy nutrient content in food crops.

Overall, understanding the role of soil microorganisms in nutrient cycling is essential for improving soil quality and increasing the nutrient content of food crops. By promoting the growth and activity of beneficial soil microorganisms through sustainable agricultural practices, we can ensure a healthy and nutrient-rich harvest for future generations.

CHAPTER 18

IMPACT OF SOIL MICROBIOLOGY ON PLANT HEALTH AND NUTRIENT CONTENT

Soil microbiology plays a crucial role in determining the health and nutrient content of plants. The complex interactions between soil microorganisms and plant roots influence the availability of essential nutrients such as nitrogen, phosphorus, and potassium. Beneficial soil microbes, such as mycorrhizal fungi and rhizobacteria, form symbiotic relationships with plants, enhancing their ability to absorb nutrients from the soil.

Research has shown that soil pH levels also play a significant role in determining the nutrient content of food crops. Soils with pH levels that are too high or too low can limit the availability of certain nutrients, leading to deficiencies in plants. Understanding and maintaining the optimal pH levels in soil is crucial for ensuring the nutrient content of food crops (Andrews et al., 2016).

Organic farming practices, which promote the use of natural fertilizers and avoid synthetic chemicals, have been shown to improve soil quality and nutrient content in food. In contrast, conventional farming practices that rely heavily on chemical inputs can degrade soil health over time, leading to nutrient deficiencies in crops.

Soil erosion is another factor that can impact the nutrient content of crops. Eroded soils are often depleted of essential nutrients, leading to lower nutrient content in food products. Implementing soil conservation practices, such as cover cropping and contour plowing, can help prevent soil erosion and maintain nutrient levels in the soil.

The use of compost and other soil amendments is another effective way to improve nutrient content in food crops. These organic materials provide essential nutrients to plants and improve soil structure, promoting the growth of beneficial soil microorganisms.

Maintaining soil health is essential for ensuring the nutrient content of food crops. By understanding the role of soil microbiology, pH levels, farming practices, erosion, and soil amendments, we can work towards producing nutrient-rich harvests that benefit both human health and the environment.

CHAPTER 19

HARNESSING SOIL MICROBES FOR SUSTAINABLE AGRICULTURE PRACTICES

Contents

Soil microbes play a crucial role in the health and fertility of the soil, ultimately impacting the nutrient content of food crops. By understanding the importance of these tiny organisms, farmers and researchers can implement sustainable agriculture practices that not only improve soil quality but also enhance the nutritional value of the food we consume.

One of the key ways in which soil microbes benefit agriculture is through their ability to break down organic matter and release essential nutrients for plant uptake. This process, known as mineralization, helps to replenish the soil with vital nutrients such as nitrogen, phosphorus, and potassium. By harnessing the power of soil microbes, farmers can reduce their reliance on synthetic fertilizers, leading to more sustainable farming practices.

19.1 ROOT RELATIONSHIPS FOR PLANT HEALTH

Furthermore, soil microbes contribute to the overall health of plants by forming symbiotic relationships with their roots. Mycorrhizal fungi, for example, can increase the surface area of plant roots, allowing for greater nutrient absorption. This symbiosis not only benefits the plants themselves but also enhances the nutrient content of the food they produce.

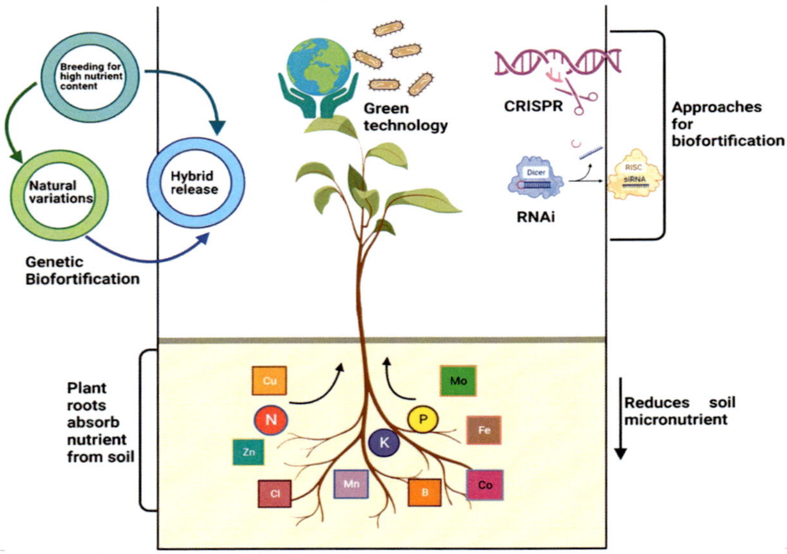

Figure 19.1. *Biofortification. Schematic illustration of micronutrients absorption in plants and different methods of biofortification. The absorption of micronutrients by plants results in the depletion of the soil micronutrients. Plants are deficient in micronutrients that can be biofortified through various approaches.*

Source: Mandal et al. (2023).

In order to promote the growth and activity of beneficial soil microbes, farmers can implement practices such as crop rotation, cover cropping, and the use of compost and other organic amendments. These methods not only improve soil health but also increase the availability of nutrients for plant uptake.

Overall, harnessing soil microbes for sustainable agriculture practices is essential for maintaining soil quality and nutrient content in food crops. By prioritizing the health of the soil and its microbial inhabitants, farmers can ensure a nutrient-rich harvest for years to come (Figure 19.1).

CHAPTER 20

MICROBIAL ENHANCEMENT OF PLANT NUTRIENT ACQUISITION

Contents

Plants usually cohabit with a plethora of soil microbes, some of which are beneficial in that they can support plant growth and/or strengthen plant resistance to one or multiple stress conditions, including nutrient deficiency (Kloepper et al. 1980; Lugtenberg and Kamilova 2009; Morcillo and Manzanera, 2021). Within the rhizosphere where root exudates give an environment enriched with organic nutrients, beneficial microbes may be free-living or directly connected with plants as epiphytes or endophytes; Especially, plant-associated fungi extend mycelia to additional soil beyond the rhizosphere and can thus reach for more nutrients to support the association. These beneficial microbes give an environment-friendly approach for enhancing plant nutrition, either as supporters or alternatives to chemical fertilizers. Some of the beneficial microbes, including rhizobia that fix atmospheric nitrogen, are well-known and have been applied in agriculture, while many others are yet to be characterized and exploited. In the following section the way in which beneficial microbes improve plant acquisition of the macronutrients N, P, K, Mg, and S as well as the micronutrients Fe, Mn, Cu, Zn, B, Mo, and Ni is discussed. Additionally, the works that investigated nutrition-dependent plant-microbe interactions, which demonstrate the plant's initiative in determining the plant-microbe association for optimized benefits including P nutrition, are discussed.

20.1. NITROGEN (N)

Microbes play paramount roles in plant acquisition of ammonium (NH_4^+) and nitrate (NO_3^-). Diazotrophs are N_2-fixing bacteria and archaea with nitrogenase that catalyzes the lowering of atomospheric N_2 to NH_3, which is then ionized to NH_4^+ favoring low environmental pH. Ammonium is also created by bacterial or fungal decomposition of organic debris and can be oxidized by nitrifying bacteria to nitric oxides and nitrates. Some N_2-fixing bacteria are free-living in soil, while others are plant-associated or need symbiosis with plants to fix N_2. Symbiotic N_2 fixation takes place in root nodules, where the specialized environment guarantees the bacterial fixation of N_2 and the bacteroids receive plant-derived organic nutrients in exchange of NH_3 (Mus et al., 2016, Roy et al., 2020). Plants capable of forming nodules mostly are part of the family of Leguminosae. These leguminous plants host species-specific symbiosis with rhizobia from the Rhizobiaceae family (Andrews and Andrews 2017). Additionally, rhizobia-legume symbiosis, N_2-fixing nodulation also take place in *Parasponia* species that host rhizobia and in actinorhizal plants that host the actinomycete *Frankia* (Svistoonoff et al., 2014, van Velzen et al., 2018).

Plant N acquisition can also be made better by non-nodulating diazotrophs. For example, inoculation with the nitrogen-fixing cyanobacterium *Nostoc punctiforme* reported detectable nitrogenase activity in rice roots and promotion of plant growth under nitrogen deficiency (Álvarez et al., 2020). *Azospirillum brasilense* strain HM053, which fixes nitrogen and excretes NH_4^+ constitutively, was able to colonize the surface of wheat roots and substantially augment the plant dry weight (Santos et al., 2017). Furthermore, horizontal transfer of the nitrogen fixation (*nif*) genes can enable the capacity of nitrogen fixation in non-diazotrophic microbes (Dixon and Postgate, 1971, Jing et al., 2020). For example, genetic engineering of the *nif* genes from a *Pseudomonas stutzeri* strain into a *Pseudomonas protegens* strain enables the latter the ability of nitrogen fixation and thus causing plant growth-promotion under nitrogen deficiency (Jing et al., 2020).

20.2. PHOSPHORUS (P)

Most natural P in soil is in insoluble form and hence unavailable to plants (Richardson and Simpson, 2011). A wide diversity of bacteria and fungi species are able to solubilize inorganic P and/or mineralizing organic P, thereby releasing bioavailable P that can be easily taken up by plants. Strains from the bacteria genera *Pseudomonas*, *Bacillus* and *Rhizobium* are among the most powerful P-solubilizing microbes (Rodríguez and Fraga, 1999). Immobilized inorganic P can be released by microbe-secreted organic acids, such as gluconic acid and citric acid. Anions from these organic acids chelate cations that otherwise precipitate P-containing anions (Whitelaw et al., 1999), at the same time acidification of the soil environment liberates P from precipitates via the creation of soluble hydrogen (HPO_4^{2-}) or dihydrogen ($H_2PO_4^-$) phosphates (Sharma et al., 2013). Protons that lead to acidification may also be released through H^+-ATPase, via transporter-assisted cation exchange across the microbial membranes, or in the form of inorganic acids (Alori et al., 2017). Microbial mineralization of organic P is mostly catalyzed by non-specific acid phosphatases that dephosphorylate the phosphor-ester or phosphoanhydride bonds of organic molecules (Nannipieri et al., 2011), in addition to by phytases that specifically hydrolyze phytate (*myo*-inositol hexakisphosphate) in liberating the orthophosphate anion (Mitchell et al., 1997, Lim et al., 2007). The hydrolysis of phytate by phytases produces *myo*-inositol (Mitchell et al., 1997), which is a frequent component of root exudates in many plant species (Vílchez et al., 2020). Phytate is the most abundant organic phosphorus compound in soil (Lim et al., 2007), while *myo*-inositol attracts *Bacillus megaterium* spp. that is well-known for strong P-solubilizing ability (Vílchez et al., 2020). Hence it is an open question whether the plant exudation of *myo*-inositol has a function in attracting *B. megaterium* spp. for bacteria-assisted P acquisition from the hydrolysis of phytate. An arbuscular mycorrhiza (AM) (plural *mycorrhizae*) is a form of mycorrhizain which the symbiont fungus (*AM fungi*, or AMF) enters the cortical cells of the roots of a vascular plant forming arbuscules.

Additionally, to giving more bioavailable P to plants, beneficial microbes may also stimulate plant P absorption via regulation of plant P transporters. For instance, colonization of AM fungi strongly increased the transcript level of *LePT4*, a phosphate carrier found in tomato plants, with a concomitant double increase in phosphate absorption; meanwhile a loss-of-function mutation in *LePT4* significantly hindered the AM fungi-enhanced phosphate uptake (Xu et al., 2007). *LePT4* belongs to the Pht1 family of plant phosphate transporters, which are expressed at both the root–soil interface and the symbiotic fungus–plant interface surrounding arbuscules or hyphal coil (Nagy et al., 2005). Phosphate transporter genes specifically activated in AM fungi symbiosis have been highlighted in many agricultural plant species, such as potato (*Solanum tuberosum*), rice (*Oryza sativa*), and *Medicago truncatula* (Rausch et al., 2001, Harrison et al., 2002, Paszkowski et al., 2002).

Fungi can be thought of as more effective tools compared to bacteria in promoting plant P acquisition, since the former release greater amounts of organic acids, reach more resources, and are able to transmit phosphate to the symbiotic plant cells (Sharma et al., 2013). Nevertheless, a combination of P-solubilizing bacteria and fungi can result in better effects than bacteria or fungi alone. For example, a combinational inoculation with AM fungi and beneficial rhizobacteria to durum wheat increased transcript levels of the P transporters Pht1 and PT2–1 as well as the aboveground P contents; in contrast, AM fungi alone induced Pht2 gene expression and needed an extra supply of organic N to similarly increase plant P contents (Saia et al., 2015). It has been demonstrated that the extraradical hyphae of AM fungi is able to transport P-solubilizing bacteria to organic P patches and enhance organic P mineralization both under *in vitro* culture and soil conditions, in a form that hyphal exudates are required as the energy source for transferring the bacteria (Jiang et al., 2021), thus highlighting the advantage of utilizing fungi and bacteria in combinations for enhancing P acquisition in plants.

20.3. POTASSIUM (K)

In most soils, the solution and exchangeable K constitute only a few percent of the total soil K (Brady et al., 2008). Many bacterial and fungal strains have been documented as capable of releasing K from silicate minerals such as mica, biotite, and orthoclase (Meena et al., 2014, Masood and Bano, 2016). These microbes enhance mineral weathering under *in vitro* conditions and increase K solubilization in soil, mainly through acidolysis, chelation and exchange reactions, in which microbe-secreted organic acids play an important role (Masood and Bano, 2016, Meena et al., 2014). K enrichment resulted from AM fungi colonization has been demonstrated in different plant organs such as

Zea mays root steles, *Pelargonium peltatum* shoots, and *Lactuca sativa* leaves (Perner et al., 2007, Baslam et al., 2013).

Many of the K-solubilizing microbes are saprophytic strains (Meena et al., 2014). Colonization by *Clitopilus hobsonii*, a common soil saprophytic fungus, increased plant growth and facilitated K uptake in the American sweetgum (*Liquidambar styraciflua*), particularly under K limitation conditions (Peng et al., 2021). Gene expression of ChACU, a putative K^+ transporter in *C. hobsonii*, were detected in the colonized roots together with two other similar genes and were up-regulated by low K^+ conditions; however, the expression of plant genes related to K^+ uptake was not altered by the symbiosis with *C. hobsonii* (Peng et al., 2021). Thus the *C. hobsonii* K^+ transporters appear to play a more active role than the sweetgum K^+ transporters in the symbiotic supply of K. In contrast, fungal induction of plant putative K^+ transporter genes was observed in perennial ryegrass (*Lolium perenne*) colonized by *Aspergillus aculeatus*, concomitant with higher levels of soluble K contents (Li et al., 2021). In another symbiosis example, the ectomycorrhizal fungus *Hebeloma cylindrosporum* significantly improved the K nutrition of pine (*Pinus pinaster*) plants under potassium-limiting conditions, and the fungal tandem-pore outward-rectifying K^+ (TOK) channels are likely responsible for releasing K^+ from the fungal Hartig net to the plant (Guerrero-Galán et al., 2018). The efficacy of microbial enhancement of plant K nutrition can be improved by co-application with K-containing minerals, as shown in perennial ryegrass treated with *A. aculeatus* plus K-containing feldspar compared with plants treated with the fungus alone (Li et al., 2021).

20.4. SULFUR (S)

Plant sulfur nutrition is dependent mostly on the absorption of inorganic sulfate (SO_4^{2-}), whereas over 95% of the soil sulfur in solid used for agriculture is present in the form of sulfate esters or as carbon-bonded sulfur in the forms of sulfonates and amino acid sulfur (Kertesz and Mirleau, 2004, Zhao et al., 2006). Various bacterial and fungal species in the rhizosphere are able to release S from sulfate esters, as an outcome of the mineralization catalyzed by sulfatases, which cover arylsulfatase that cleaves the O-S bond in aromatic sulfate-esters and alkylsulfatase that cleaves the C-O bond in aliphatic sulfate-esters (Kertesz, 2000). As opposed to the mineralization of sulfate esters, releasing sulfur from sulfonates is catalyzed by a multicomponent mono-oxygenase enzyme complex encoded in the *ssu* gene cluster (Eichhorn et al., 1999), which is present in bacteria but not fungi (Gahan and Schmalenberger, 2015). The ability of releasing S from aliphatic sulfonates is widespread in soil bacteria (King and Quinn, 1997), whereas liberating S from aromatic sulfonates needs the *asfRABC* gene cluster in addition to the *ssu* gene cluster (Vermeij et al., 1999). Aliphatic and aromatic sulfonates combined account for 30–70% soil S (Zhao et al., 2006). It was proposed that aromatic sulfonates are more influential than aliphatic sulfonates in microbe-assisted plant sulfur nutrition, since bacterial ability in using the former but not the latter has been connected with plant growth-promotion (Kertesz and Mirleau, 2004, Gahan and Schmalenberger, 2015). Even though fungi not able to mineralize sulfonates, a few saprotrophic fungi may facilitate bacterial mineralization of sulfonates through depolymerizing polymeric sulfonates into monomeric or oligomeric sulfonates (Gahan and Schmalenberger, 2015, Kertesz et al., 2007).

As well as increasing the supply of SO_4^{2-}, soil microbes may additionally enhance plant sulfur nutrition via transcriptional regulation of plant sulfate assimilation pathway. In *Arabidopsis thaliana* exposed to the volatiles organic compounds (VOCs) released from *Bacillus amyloliquefaciens* GB03, transcript levels of several genes, such as ATP sulfurylase (ATPS) and adenosine 5′ – phosphosulfate reductase (APR), in the sulfate assimilation pathway were up-regulated; concomitantly, VOC-enhanced plant sulfate uptake was seen with radio-labeled sulfate within 30 min (Aziz et al., 2016). In *Lotus japonicas* colonized by the AM fungus *R. irregularis*, gene expression of the sulfate transporter LjSultr1;2 was found in arbuscule-containing cells and up-regulated by either the fungal colonization or S starvation (Giovannetti et al., 2014). Dysfunction of LjSultr1;2 lowered plant constitutive sulfur uptake, while the fungal colonization augmented plant sulfate contents under sulfur starvation (Giovannetti et al., 2014), and in this way supporting the importance of transcriptional regulation in the microbe-assisted plant sulfur nutrition.

Differently from other microbes that strengthen plant sulfur absorption in roots, *Bacillus* sp. B55 is able to augment plant sulfur uptake in the air (Meldau et al., 2013). With radio-labeled sulfur S added into the bacteria growth medium, B55 VOCs was demonstrated to transmit sulfur to *Nicotiana attenuata* plants, which were grown in a medium physically separated from the bacteria growth medium, and rescued plant growth delay originating from sulfur starvation. Dimethyl disulfide (DMDS) was highlighted as the major S-containing component in B55 VOCs and responsible for the VOC-enhanced plant S nutrition (Meldau et al., 2013). Sulfur in SO_4^{2-} is in an oxidative state that requires an energy-consuming reduction process for biological assimilation (Takahashi et al., 2011), whereas sulfur in DMDS is in a chemically reduced form, which may credit DMDS as an energy-saving form of sulfur nutrition to plants. Consistent with this hypothesis, sulfur uptake as well as methionine biosynthesis and recycling in plants were transcriptionally repressed by DMDS, reducing demand for SO_4^{2-} (Meldau et al., 2013). DMDS is a frequent component in microbial VOCs; simultaneously other microbes may produce

VOCs with high amounts of other sulfur-containing compounds including dimethyl sulfide and dimethyl trisulfide (Kanchiswamy et al., 2015). Thus, VOC-mediated sulfur assimilation in the air might be a common mechanism for microbe-enhanced plant S nutrition.

20.5. MAGNESIUM (MG)

Plants grown in acidic and sandy soils often suffer from magnesium deficiency because of leaching and antagonism by other cations (Hermans et al., 2010). Some examples have been identified for microbe-mediated alleviation of plant Mg deficiency. In *A. thaliana* particularly under magnesium deficiency condition, the plant magnesium nutrition was made better by the colonization of *Piriformospora indica*, a root endophytic fungus that possesses the magnesium transporter PiMgT1 (Prasad et al., 2019). Inoculation of AM fungus *Funneliformis mosseae* to *Poncirus trifoliata* seedlings produced enhanced development promotion and suppression of magnesium deficiency symptoms; the lowering in magnesium deficiency symptoms may be due to the fungus-induced increases in whole-plant soluble protein amounts and in the activity of root catalase and superoxide dismutase (Zhang et al., 2015). Interesting, the colonization of *F. mosseae* to *P. trifoliata* roots was notably higher under magnesium-deficiency than under magnesium-sufficiency (Zhang et al., 2015), while sweet potato and onion leaves sprayed with elevated levels of $MgSO_4$ showed strongly inhibited root AM fungus colonization and sporulation in aeroponic and sand culture (Jarstfer et al., 1998), demonstrating a negative correlation between plant magnesium availability and AM fungi colonization to roots.

20.6. IRON (FE)

Plants utilize various strategies to augment iron mobility in the rhizosphere. Graminaceous monocots liberate siderophores to chelate ferric iron (Fe^{3+}) prior to its root absorption, while non-graminaceous monocots and dicots deploy a combined strategy including rhizosphere acidification, reduction of Fe^{3+} to ferrous iron (Fe^{2+}) by plasma membrane ferric reductase, and transporter-mediated Fe^{2+} import into roots (Curie and Briat, 2003). These strategies participate in microbial strengthening of plant iron uptake.

Microbial siderophores in their iron-bound forms may be directly taken up by plants. Hydroxymates and catechols are the main functional groups in the iron-siderophore binding (Rajkumar et al., 2010). Siderophores produced from various microbes may be characterized by their structural variations, for example, siderophores in the form of rhizobactin (catechol-containing) and citrate (neither a catechol nor a hydroxamate) are secreted from *Rhizobium meliloti* and *Bradyrhizobium japonicum*, correspondingly (Smith et al., 1985, Guerinot et al., 1990). As well as providing microbial siderophores, some microbes additionally stimulate plant production of phytosiderophores. In sorghum (*Sorghum bicolor*) under iron deficiency, gene expression of SbDMAS2 (deoxymugineic acid synthase 2), SbNAS2 (nicotianamine synthase 2), and SbYS1 (Fe-phytosiderophore transporter yellow stripe) was induced by AM fungus, accompanying noticeable augments in phytosiderophore release and iron uptake in the plant (Prity et al., 2020).

Acidification ameliorates Fe^{3+} movement, thus, rhizosphere acidification stimulated or directly contributed by microbes improves plant iron uptake. Organic acids commonly exist in microbial extracellular metabolites. For example, the VOCs from *B. amyloliquefaciens* GB03 have glyoxylic acid, 3-methyl-butanoic acid and diethyl acetic acid (Farag et al., 2006), which directly lead to rhizosphere acidification (Zhang et al., 2009). GB03 VOCs also promote plant acidification of the rhizosphere, as demonstrated by VOC-exposed Arabidopsis that were moved to new growth medium with no VOCs anymore (Zhang et al., 2009). Noteworthy, Arabidopsis under iron-sufficient conditions responded to GB03 VOCs with systemic iron-deficiency responses, including transcriptional induction of the Fe^{3+}reductase FRO2 and the Fe^{2+} transporter IRT1, in addition to elevated enzyme activity of FRO2, as well as to the stimulation of rhizosphere acidification by the plant (Zhang et al., 2009). Thus, iron levels increased in the VOC-exposed plants (Zhang et al., 2009), in supporting plant production of higher levels of Fe-containing photosynthetic apparatus (Zhang et al., 2009, Zhang et al., 2008). The phenotype and/or mechanism of GB03 VOC-enhanced plant iron uptake were similarly seen in many later studies, including investigations on other plants species, VOCs from other microbes, and soil inoculation with GB03 or other beneficial microbes (del Carmen Orozco-Mosqueda et al., 2013a, del Carmen Orozco-Mosqueda et al., 2013b, Castulo-Rubio et al., 2015, Freitas et al., 2015, Zamioudis et al., 2015, Zhou et al., 2016, Martínez-Medina et al., 2017, Hernández-Calderón et al., 2018, Montejano-Ramírez et al., 2020, Kong et al., 2021). Thus, microbial control of plant iron-deficiency responses plays a key function in microbe-enhanced plant iron acquisition.

The volatile compound *N,N*-dimethylhexadecylamine (DMHDA), which is often generated by various bacteria strains belonging to *Arthrobacter agilis*, *Sinorhizobium meliloti*, or *Pseudomonas fluorescens*, was demonstrated to lead to plant iron-deficiency responses (del Carmen Orozco-Mosqueda et al., 2013a, Castulo-Rubio et al., 2015, Montejano-Ramírez et al., 2020). GB03 VOCs' effects on plant iron-acquisition needs transcriptional induction of FIT1 (Zhang et al., 2009), a master regulator of iron-deficiency responses kept in plants (Riaz and Guerinot, 2021). The bioactive VOC component(s) responsible for this essential controlling step along with its mechanism are not yet fully understood. Even though identifying the bioactive component or combination from natural microbial VOC blends has been difficult, the positive impacts of microbial VOCs on plant resistance to biotic and abiotic stress have been adding up (Farag et al., 2013, Liu and Zhang, 2015, Weisskopf et al., 2021), making these microbial metabolites key components for the development of new tools for agricultural applications.

20.7. COPPER (CU)

AM fungi are able to augment plant fitness under copper deficient and toxic conditions (Ferrol et al., 2016). In white clover (*Trifolium repens* L.) plants colonized by *Glomus mosseae*, the amelioration in plant copper nutrition was assumed to come from extraradical mycelia that went out of the root into the surrounding environment for additional nutrients (Li et al., 1991). Likewise, *G. mosseae* colonization increased copper contents in the shoots of cucumber (*Cucumis sativus*), owing to extraradical mycelia-dependent nutrient absorption (Lee and George, 2005). In the mycorrhizalwhite clover, the fungal role to plant copper absorption was independent of phosphorus availability; nevertheless, higher exogenous P supplies from soil changed the distribution patterns of the copper contents provided by *G. mosseae*, with a reduction and an increase in the roots and the shoots, correspondingly (Li et al., 1991). Likewise, in the mycorrhizal cucumber plants, large amounts of phosphorus supply to hyphae caused a reduction in root copper concentrations (Lee and George, 2005), and in this way bringing attention to the connection among different nutrient betterment in mycorrhizal plants.

20.8. MANGANESE (MN)

Plants absorb manganese exclusively in its divalent form (Mn^{2+}). Manganese is minimally available for plant absorption in well-aerated calcareous soils, since alkaline and oxidative conditions favor the development of water-insoluble manganese oxides; meanwhile Mn^{2+} soil dressing is quite ineffective to alleviate deficiency because of rapid oxidation of Mn^{2+} (Andresen et al., 2018). Because manganese availability in soils is established by pH, microbe-mediated rhizosphere acidification would increase plant manganese acquisition. Examples of bacterial mobilization of manganese can be seen in some acidophilic bacteria strains, including *Acinetobacter* sp. and *Lysinibacillus* sp., which were utilized for manganese bioleaching from ores or mining waste water (Sanket et al., 2017, Ghosh et al., 2018). Likewise, manganese solubilizing fungal strains, including *Aspergillus terreus* and *Penicillium daleae*, have been isolated from low-grade manganese mine tailings, and their manganese solubilizing ability was assumed to come from to the mycelia production of organic acids including oxalic acid, citric acid, maleic acid and gluconic acid (Mohanty et al., 2017).

Because of its low selectivity, the Fe^{2+} importer IRT1 is additionally one of the manganese transporters in plants (Barberon et al., 2014). Furthermore, the capacity of both iron and manganese in soil are established by pH and redox potential. Hence, plant manganese uptake may be strengthened by microbes that are able to trigger plant iron deficiency responses. In a calcareous soil, the rhizosphere microbiome of maize (*Z. mays*) harbored populations of Fe^{3+}-reducing microbes and Mn^{4+}-reducing microbes, with the latter being one magnitude greater than the former (Kothari et al., 1991).

20.9. ZINC (ZN)

Zn deficiency often limits the yield and quality of staple food crops. It has been suggested that approximately 50% of the rice-growing soils are zinc-deficient, and that zinc fertilization by employing the soluble $ZnSO_4$ usually results in minimal fertilizer use efficiency (1–5%) because of the zinc immobilization in the soil (Krithika and Balachandar 2016). Zinc-mobilizing microbes provide different tools for strengthening plant zinc acquisition. In rice (*O. sativa*) seedlings, similar effects on plant zinc accumulation were seen between the $ZnSO_4$ fertilization and the utilization of insoluble zinc oxide bound with zinc-mobilizing bacteria (Krithika and Balachandar 2016). In a different study with natural soil deficient in zinc, utilization of Zn-mobilizing bacteria produced better effects in augmenting plant zinc contents than zinc fertilization at the rate of 2.5 mg Zn as $ZnSO_4$ $7H_2O$ per kg soil (Vaid et al. 2014). A diversity of bacteria including *Pseudomonas* and *Bacillus* strains are known to increase plant growth with greater amounts of zinc, and the microbial movement of zinc were attributed to various mechanisms controlled by rhizosphere acidification, sequestration by siderophores or anions from organic acids, and oxido-reductive systems on cell membranes (Kamran et al., 2017). Fungi-assisted plant Zn acquisition is also apparent. In mycorrhizal bread wheat (*Triticum aestivum*) and barley (*Hordeum vulgare*), the fungal contribution of plant zinc uptake was established as up to 24.3% and 12.7% of total above-ground zinc in wheat and barley, respectively (Coccina et al. 2019). In mycorrhizal maize, the plant physiological function including photosynthesis in zinc deficient soils was improved with greater nutrient absorption (Saboor et al. 2021).

Additionally, to giving plants mycorrhizal zinc supply, fungi and bacteria can also control plant zinc transporters. Inoculation of barley with the AM fungus *R. irregularis* resulted in strong gene up-regulation of HvZIP13, a Zn transporter gene that is started by minimal zinc availability, in tandem with improved grain and straw zinc concentrations (Watts-Williams and Cavagnaro, 2018). Likewise, *R. irregularis* inoculation augmented plant tolerance to soil zinc deficiency in *M. truncatula*; concomitantly, the fungus set off gene expression of MtZIP5 that is inducible by zinc deficiency (Nguyen et al. 2019).

Of note, although AM fungi are able to improve plant zinc nutrition under minimal soil zinc conditions, fungal protection of plants from too much zinc accumulation at elevated zinc availability have also been reported in different plant species including red clover (*Trifolium pretense*) and tomato (Li and Christie 2001, Watts-Williams et al. 2013, Nguyen et al. 2019). These dual plant-beneficial effects make AM fungi fantastic options as microbial tools for optimizing crop zinc acquisition.

20.10. MOLYBDENUM (MO)

Plants absorb molybdenum in molybdate (MoO_4^{2-}) form. Apart from copper, molybdenum is the lowest abundant essential micronutrient present in the majority of plants (Kaiser et al., 2005). Molybdenum deficiencies are thought of as unusual in the majority of agricultural cropping regions, meanwhile its fertilization via foliar sprays may effectively supplement internal molybdenum and save the activity of molybdenum-dependent enzymes (Kaiser et al., 2005). Even so, plant molybdenum uptake can be enhanced by specific microbes. In sweet sorghum grown in an agricultural soil spiked with various concentrations of MoS_2, AM fungal colonization by a *Claroideoglomus etunicatum* strain strongly increased plant molybdenum concentrations in both shoots and roots (Shi et al., 2020). In maize plants growing in soil supplemented with various amount of $(NH_4)_2MoO_4$, the same *C. etunicatum* strain also strengthened plant molybdenum concentrations in shoots and roots, with lower levels in the shoot-to-root molybdenum ratio when molybdenum was supplemented at the amounts deemed as moderate and severe contamination (Shi et al., 2018).

20.11. NICKEL (NI)

Nickel is needed in higher plants only by the enzyme urease. The mechanism of its uptake by plants is greatly unknown (Andresen et al., 2018). This micronutrient is usually present in fertilized soil in the form of additives to fertilizers. As an example, rock phosphate, being a raw material for phosphatic fertilizers, is known to contain Ni ranging between 16.8 to 50.4 mg kg^{-1} (Chauhan et al., 2008). Ni is greatly known as a common heavy metal contaminant in soil, air, and water. Thus, the study of microbes for improving plant Ni homeostasis has been targeting microbe-induced plant stress-alleviation, which is often regulated by lowering heavy metal uptake and/or augmenting plant tolerance to the metal ion toxicity.

CHAPTER 21

DIFFERENCES IN SOIL QUALITY BETWEEN ORGANIC AND CONVENTIONAL FARMING

Contents

When it comes to the quality of soil used in farming, there are significant differences between organic and conventional practices. Organic farming focuses on building healthy soil by using natural methods such as composting, cover cropping, and crop rotation. On the other hand, conventional farming often relies on synthetic fertilizers and pesticides to boost crop yields.

One of the key differences between organic and conventional farming is the impact on soil microbiology. Organic farming practices promote a diverse and healthy soil microbiome, which plays a crucial role in nutrient availability for plants. Conventional farming, on the other hand, can disrupt the balance of soil microorganisms through the use of chemical inputs.

Another important factor to consider is soil erosion, which can have a direct impact on the nutrient content of crops. Organic farming practices, such as no-till farming and the use of cover crops, can help prevent soil erosion and preserve nutrient-rich topsoil. Conventional farming, which often involves intensive tillage, is more likely to contribute to soil erosion and nutrient loss (Reganold et al, 2010).

In terms of soil pH levels, organic farming tends to result in slightly more acidic soils compared to conventional practices. While this can have implications for nutrient availability, organic farmers can make adjustments by using natural amendments such as lime or compost.

21.1 ORGANIC VS CONVENTIONAL FARMING

Overall, the choice between organic and conventional farming practices can have a significant impact on soil quality and ultimately, the nutrient content of food. By understanding the differences between these approaches and implementing sustainable agriculture practices, we can work towards maintaining healthy soil and producing nutrient-rich crops for a healthier food supply.

The global population is currently growing at an exponential speed and as the demand for food production grows, huge pressure and demands will be positioned on all natural resources including soil. Conventional methods of farming heavily cover the extensive use of synthetic fertilizers and agrochemicals with heavy dependency on tillage practices which will keep being the main method of production and the penalty will be shown in terms of lowered soil quality, impacting the soil's capacity for production (Gomiero et al., 2011). On the other hand, organic farming attempts to copy or use natural processes as guidelines that are likely to improve soil and plant health while maintaining soil and water resources (Gomiero et al., 2011). These intensive farming practices have produced a reduction in soil organic matter impacting the soil physical, chemical and biological properties.

The lowering of soil quality because of a reduced amount of organic matter is creating a significant risk to sustainability of present agricultural systems which are fully relying on agro-chemicals and high energy inputs. These inputs represent a strong negative effect for soil and water, leading to degradation in form of erosion, alkalinity, salinity, acidification, water logging, macro and micronutrients deficiencies that ultimately impact soil quality and crop productivity (Lopes et al., 2011). Additionally, strong use of fertilizers and pesticides lead to the increase of toxic elements that are resistant to degradation in the soil. It is well understood that overall, conventional farming causes degradation of soils, and certainly will lead to the creation of problems in meeting sustainability objectives in terms of agricultural production. These obstacles to sustainability will be sped up by impacting natural weather phenomena including longer periods of drought, which may bepart of the ongoing climate change (Montgomery, 2007a).

Additionally, leaching losses of nitrate lead to eutrophication of surface waters and contamination of groundwater. When soil applied pesticides do not reach the target, they impact surrounding ecosystems via leaching or aerial drift impacting the diversity and quantity of non targeted microorganisms having negative effects on ecosystem processes and trophic interactions (Pimentel and Edwards, 1982). There is a mounting acknowledgment among agricultural scientists that although the conventional farming practices have led to achieve the target of increased food production, it was at the cost of food quality, environmental deterioration and degradation of natural resources. Thus, there is an increasing worry regarding the environmental, economic and social impacts of chemical-dependent conventional farming systems that have caused the expert community to look for alternative systems that can turn agriculture more sustainable and profitable.

CHAPTER 22

SUSTAINABLE AGRICULTURE PRACTICES FOR MAINTAINING SOIL QUALITY AND NUTRIENT CONTENT

Contents

22.1. PRINCIPLES OF SUSTAINABLE AGRICULTURE FOR SOIL HEALTH

In order to achieve sustainable agriculture practices for soil health, it is important to understand the principles that guide successful farming techniques. One of the key principles is the importance of maintaining soil quality in order to ensure nutrient-rich harvests. Soil quality directly impacts the nutrient content of food crops, making it essential for food scientists, university students, researchers, and policy makers to prioritize the health of the soil.

Soil pH levels play a crucial role in determining the nutrient content of food. Understanding how different pH levels can affect nutrient availability is vital for maximizing the nutritional value of crops. Organic farming practices have been shown to have a positive impact on soil quality compared to conventional methods. By avoiding synthetic chemicals and focusing on natural inputs, organic farmers can improve soil health and nutrient content in their produce.

Soil erosion is another factor that can significantly impact the nutrient content of crops. Implementing erosion control measures is essential for maintaining soil quality and preserving the nutrient levels in food products. Additionally, soil microbiology plays a key role in nutrient availability for plants. By promoting healthy soil microbial communities, farmers can ensure that their crops have access to essential nutrients (Philippor et al., 2013). The use of compost and other soil amendments is a sustainable way to improve nutrient content in food. By recycling organic matter back into the soil, farmers can enhance soil health and increase the nutrient levels in their crops. Sustainable agriculture practices, such as crop rotation and cover cropping, are also effective ways to maintain soil quality and nutrient content in food products.

Overall, understanding the relationship between soil health, plant health, and nutrient content is essential for promoting sustainable agriculture practices. By prioritizing soil quality and implementing strategies to improve nutrient availability, farmers can ensure that their crops are nutrient-rich and contribute to a healthy food system. Soil testing methods are important tools for assessing nutrient levels in food crops and guiding farming practices towards optimal nutrition. By following these principles of sustainable agriculture, we can work towards a future where nutrient-rich harvests are the norm.

22.2. IMPLEMENTING SUSTAINABLE PRACTICES TO ENHANCE NUTRIENT CONTENT IN FOOD

Implementing sustainable practices is crucial in enhancing the nutrient content in food. By focusing on soil quality, we can ensure that the food we consume is rich in essential nutrients.

Soil pH levels play a significant role in determining the nutrient content of crops. It is important to monitor and adjust pH levels to optimize nutrient availability for plants. Organic farming practices have been shown to improve soil quality and increase nutrient content in food compared to conventional farming methods. By avoiding the use of synthetic chemicals, organic farmers maintain healthier soil that is better able to support nutrient-rich crops.

Soil erosion can deplete the soil of essential nutrients, leading to lower nutrient content in crops. Implementing erosion control measures, such as cover cropping and contour plowing, can help maintain soil health and preserve nutrient levels in food.

Soil microbiology also plays a key role in nutrient availability for plants. Beneficial soil microbes help break down organic matter and release nutrients for plant uptake. By promoting healthy soil microbiology through practices like crop rotation and minimal tillage, we can enhance the nutrient content of food.

The use of compost and other soil amendments is another effective way to improve nutrient content in food. These organic materials provide essential nutrients to the soil, supporting healthy plant growth and enhancing the nutrient levels in crops.

Sustainable agriculture practices, such as crop diversification and integrated pest management, are essential for maintaining soil quality and nutrient content in food. By prioritizing soil health, we can ensure that our food is not only delicious but also nutrient-rich.

Soil testing methods are crucial for assessing nutrient levels in food crops. By regularly testing the soil and adjusting nutrient inputs as needed, farmers can optimize nutrient content in their crops.

Overall, the relationship between soil health, plant health, and nutrient content in food products is undeniable. By implementing sustainable practices that prioritize soil quality, we can enhance the nutrient content of our food and promote a healthier diet for all.

22.3. CHALLENGES AND OPPORTUNITIES FOR SUSTAINABLE AGRICULTURE IN THE FUTURE

As we look towards the future of agriculture, it is clear that there are both challenges and opportunities that lie ahead in ensuring sustainable practices that promote soil quality and nutrient-rich food production. One of the key challenges that we face is the degradation of soil quality due to erosion, overuse of chemical fertilizers, and other unsustainable farming practices. This can have a direct impact on the nutrient content of the food that we produce, leading to crops that are less nutritious and potentially harmful to human health.

However, there are also opportunities for innovation and improvement in sustainable agriculture that can help address these challenges. By focusing on organic farming practices that promote soil health and biodiversity, we can improve the nutrient content of our food while also protecting the environment. One way to do this is by using compost and other soil amendments to replenish nutrients in the soil and promote a healthy microbial community that can support plant growth.

Additionally, advancements in soil testing methods can help farmers assess nutrient levels in their crops and make informed decisions about how to improve soil quality. By understanding the relationship between soil health, plant health, and nutrient content in food products, we can develop more sustainable agriculture practices that benefit both the environment and human health.

The future of sustainable agriculture will require collaboration and innovation from food scientists, university students, researchers, and policy makers. By focusing on improving soil quality, promoting organic farming practices, and utilizing soil testing methods, we can work towards producing nutrient-rich food that supports the health of both people and the planet.

Agriculture, which is the source for all of the world's food and takes up approximately 40% of the available land, is both the biggest industry in the world and a main land utilizer (Singh et al., 2019). The economic development of the world is heavily reliant on agriculture. Developing countries keep facing the heaviest pressing difficulties because of the growing population and connected food security. In the twentieth century, supporting food production was only possible via augmenting agricultural output with outside inputs such as mineral fertilizers and pesticides (Tubiello et al., 2008; Urgesa et al., 2016; Mekuria, 2018). Modern agricultural methods feeding on the green revolution have led to a noticeable growth in grain productivity at the expense of depleting natural resources. Soil fertility and environmental resilience were significantly damaged due to agriculture's externalization. Hence, it calls for different strategies that should empower farmers to use their traditional knowledge to grow more grains with less outside inputs. This practice is referred to as sustainable agriculture, which is essential.

The integrated utilization of a diversity of soil, nutrients, and pest management techniques such as manure, crop residue, mixed cropping, and crop rotations has been encouraged in sustainable agriculture systems (DeLonge et al., 2016; Mishra and Nayak, 2004; Mutyasira et al., 2018; Brock et al., 2021). By reducing soil degradation, these approaches augment soil quality, nutrient pools, climatic resilience, and ecosystem restoration, increasing the socioeconomic status of farmers. Nevertheless, turning to exclusively organic management of agroecosystems is hard because of the relatively limited supply of organic fertilizers. So as to improve soil quality and nutrient availability, integrated nutrient management (INM), which mixes increased organic inputs into the soil with a balanced application of inorganic fertilizers, is encouraged (Singh et al., 2019; Zerssa et al., 2021). Briefly said the key strategy of sustainable agriculture is that it must produce more while using fewer natural resources. Many new alternative methods, including conservation agriculture and organic farming for soil management, water-saving agronomic practices, INM, biofertilizers, and precision agriculture, have been put forward to achieve the fundamental goals of sustainable agriculture.

There is no argument that conventional agriculture (CA) has large productivity, yet it also creates environmental and social impacts of global concern (DeLonge et al., 2016; Darnhofer et al., 2010). CA has an effect on the environment via greenhouse gas emissions, water quality decline, soil erosion, livelihoods, and food security (DeLonge et al., 2016; Mishra and Nayak, 2004). CA practices are adding to the degradation of important ecological processes that maintain life on Earth. These practices are responsible for climate change, the degradation of the biosphere, the destruction of the land system, and the eutrophication of the oceans as a result of mineral fertilizers application (Brock et al., 2021; Adimassu et al., 2016). Popularizing sustainable agriculture practices among the field level via academic curricula, seminars, symposiums, and research is paramount to achieving a more promising future.

Agriculture could turn sustainable with careful modification of its components. The larger components of sustainable agriculture are achieving healthy soil and preventing erosion, using water wisely, augmenting carbon sequestration, strengthening resilience to extreme weather, and championing biodiversity. The components linked to the soil ecosystem have gained the heaviest attention due to it substantially impacting all other factors and sustainable agriculture. Utilization of organic fertilizers including compost, green manure, vermicomposting, and cover crops supported by a planned rotation of crops are utilized in sustainable agriculture to keep and strengthen soil fertility. Achieving healthy soil, augmenting carbon sequestration, and resilience to extreme weather are linked and could be achieved via these methods (Brock et al., 2021; Zerssa et al., 2021). Additionally, regionally accessible organic wastes could be turned into plant nutrients, lowering the negative impact of artificial fertilizers on sustainable soil management. Creating guidelines to reuse various wastes as plant nutrients is key to reducing the use of synthetic fertilizer. The main issues with artificial fertilizer implementation in industrialized countries are over application. Nevertheless, low-rate applications are a problem in developing countries. Land resources and agricultural production are heavily affected in both excessive and low application scenarios. Supplementing with organic fertilizer obtained from waste materials via composting is key to balancing the constraint originating from excessive and low application (Zerssa et al., 2021; Agegnehu et al., 2016). Not only does achieving healthy soil increase productivity, but it also turns the soil more drought-resilient and augments carbon sequestration. As an outcome, it might be more simple to adapt to a changing climate and lower greenhouse gas emissions (Nagothu, 2016; Song et al., 2015; Smith et al., 2019). Augmenting soil organic matter (SOM) quantities in the soils can increase water retention to a greater degree at field capacity by improving bulk density and porosity and in this way increase the plant available water capacity (PAWC) (Lal, 2020; Kane et al., 2021; Reidsma et al., 2015). The availability of SOM material may impact how much PAWC increases. Hence, building up the SOM content of soils may turn them climate-resilient. Augmenting soil resilience to drought could guarantee optimum crop yield in dry seasons (Kane et al., 2021); thus, it improves sustained yield intensification.

Because of the heavy use of synthetic chemical fertilizers and pesticides, the agricultural sector of developed countries is one of the main sources of environmental contamination (Abou-El-Hassan et al., 2017; Sun et al., 2018; Aulakh and Malhi, 2005). To augment productivity per unit area, agricultural systems worldwide utilize many chemicals including fertilizers, insecticides, and herbicides (Abou-El-Hassan et al., 2017). These chemicals and fertilizers add to a number of issues, such as soil, water, and air pollution, lowered input efficiency, lower food quality, soil degradation, missing micronutrients in the soil, toxicity to various beneficial living organisms inhabiting above and below the soil surface, and reduced production. In addition, overuse of synthetic nitrogen fertilizer may lead to large atmospheric emissions of nitrous oxide (Han et al., 2015).

The utilization of fertilizer has greatly increased globally since 1961. Worldwide data indicates that the production of synthetic nitrogen fertilizers that are given to crops has gone up approximately sevenfold, whereas the yield production of crops has augmented by almost 2.4 times (Awaad et al., 2021; Suleiman et al., 2021). Additionally, about 60% of all synthetic chemical fertilizer production was composed of nitrogen. The heavy attention given to this fertilizer significantly impacts natural soil ecosystems, which in turn affects sustainable agriculture. As an outcome, growing agricultural goods that are chemical-free and high in nutrition is paramount for both human and animal use. To stop the degradation of natural resources, the focus must be placed on producing high quality and quantity food (Seleiman et al., 2021). Generally, using excess mineral fertilizer has a negative influence on sustainable agriculture.

Application of advanced technology by small scale farmers can strengthen their income and livelihood (Adenle et al., 2019; Amede, 2015; Asfaw et al., 2012). In the majority of developing countries, agricultural practices are dominated by numerous traditional approaches. For example, in comparison to other regions, the sub-Saharan African nations, which depend mostly on rain-fed agriculture, have been negatively impacted by climate change and variability (Adela et al., 2019; Adugna et al., 2014). Because of low adoption of advanced irrigation practices such as drip irrigation and water harvesting, the area is quite susceptible to climate change and variability. Drip irrigation is a methodology that preserves water and lowers water losses via drainage. This practice was created in answer to the incentives present in the Middle East, a region that suffers from a severe water scarcity problem and which made a concerted effort to develop water-conserving technologies (Adenle et al., 2019).

Box 1: Climate Change and Its Impact on Agriculture

Climate change and variability are the most significant factors challenging sustainable agriculture (SA) worldwide (Nagothu, 2016; Reidsma et al., 2015). Climate change and variability have an impact on the cereal production system via unpredictable rainfall, flooding, and escalating mean temperature (Kotir, 2011). The majority of developing countries are strongly vulnerable to climate variability because of extremely strong dependence on rain-irrigated agriculture, low development of irrigation schemes (Nagothu, 2016; Berhane, 2018; Gezie, 2019). Vulnerability to climate variability is additionally worsened by a lack of improved technologies, missing effective early warning systems, and limited and fragmented land size (Kotir, 2011). For example, in Ethiopia, it is estimated that agricultural production will go up by 6–33% between 2030 and 2050 because of the impacts of climate change and variability (Zerssa et al., 2021; Zewdu, 2020). This will lead to food and nutrition insecurity and augment the loss of life in the country. The impacts of climate change on agriculture will depend on the speed and severity of the modification and how successfully farmers can adapt. The impact of climate change and variability is very evident in developing countries in comparison to developed countries. In order to keep and maintain sustainable agriculture and food security, developing countries must receive much more attention.

For any specific plant, the impact of higher temperature will be dependent on the plant's optimal temperature for development and reproduction. In certain areas, warming can benefit the types of plants that are usually planted there, or allow farmers to change to crops that are currently developing in warmer areas. Conversely, if the higher temperature goes over a crop's optimum temperature, yields will be reduced.

Greater CO_2 levels can affect crop yields. Certain laboratory experiments propose that higher CO_2 levels can support plant development. Nevertheless, other components, such as varying temperatures, ozone, and water and nutrient constraints, can counteract these potential increases in yield.

For instance, if temperature is above a crop's optimal level, if enough water and nutrients are not sufficient, yield increases may be lowered or reversed. Elevated CO_2 has been linked with lower protein and nitrogen amounts in alfalfa and soybean plants, resulting in a reduction of quality. Reduced grain and forage quality can decrease the ability of pasture to support grazing animals.

More extreme temperature and precipitation may hinder crops from growing. Extreme events, particularly floods and droughts, can damage crops and lower yields. For instance, in 2010 and 2012, greater nighttime temperatures impacted corn yields in the U.S. Corn Belt, and premature budding due to a warm winter caused millions in losses.

Handling drought could turn into a problem in regions where increasing summer temperatures cause soils to become drier. Although greater irrigation can be possible in some areas, in other locations water supplies may also be lowered, leaving less water accessible for irrigation when more is required.

Numerous weeds, pests, and fungi perform really well under warmer temperatures, wetter climates, and higher CO_2 levels. Currently, farmers in North America spend over $11 billion per year to control weeds, which compete with crops for light, water, and nutrients. The types and distribution of weeds and pests are likely to go up with climate change. This could lead to new challenges for farmers' crops previously unexposed to these species.

Although increasing CO_2 may stimulate plant development, it also lowers the nutritional composition of the majority of food plants. Increasing levels of atmospheric carbon dioxide lower the amounts of protein and key minerals in the majority of plant species, including wheat, soybeans, and rice. This direct impact of increasing CO_2 on the nutritional levels of plants represents a possible threat to human health. Human health is also in danger by higher pesticide use due to higher pest pressures and reductions in the efficacy of pesticides.

22.4. IMPORTANCE OF SOIL TESTING FOR UNDERSTANDING NUTRIENT AVAILABILITY

Soil testing is a crucial step in understanding the nutrient availability in the soil, which directly impacts the nutrient content of food crops. Food scientists, university students, researchers, and policy makers involved in the study of soil quality and food nutrient content must recognize the importance of soil testing for ensuring the production of nutrient-rich harvests.

One key aspect of soil testing is assessing the pH levels of the soil. Soil pH levels play a significant role in determining nutrient availability to plants. Soil that is too acidic or alkaline can limit the uptake of essential nutrients by crops, resulting in nutrient deficiencies in the harvested food products. By testing the soil pH levels, farmers can make informed decisions about the use of soil amendments to adjust the pH and improve nutrient availability for plants.

Furthermore, the choice between organic and conventional farming practices can also impact soil quality and nutrient content in food crops. Organic farming practices focus on building soil health through the use of compost and other organic amendments, which can improve the nutrient content of food products. On the other hand, conventional farming practices may rely on synthetic fertilizers that can degrade soil quality over time, leading to lower nutrient availability in crops.

In addition to soil testing, sustainable agriculture practices are essential for maintaining soil quality and nutrient content in food crops. Practices such as crop rotation, cover cropping, and reduced tillage can help prevent soil erosion and promote soil health, ultimately leading to nutrient-rich harvests.

Overall, soil testing is a critical tool for assessing nutrient levels in food crops and understanding the relationship between soil health, plant health, and food nutrient content. By implementing sustainable agriculture practices and using soil amendments strategically, farmers can improve soil quality and ensure the production of nutrient-rich food products for the benefit of consumers and the environment.

CHAPTER 23

IMPACT OF FARMING PRACTICES ON NUTRIENT CONTENT OF FOOD

Contents

Farming practices have a significant impact on the nutrient content of the food we consume. The quality of the soil in which crops are grown plays a crucial role in determining the nutritional value of the produce. Soil pH levels, for example, can affect the availability of nutrients to plants, ultimately influencing the nutrient content of the food we eat. Organic farming practices, which focus on maintaining soil health through natural means such as composting and crop rotation, have been shown to produce food with higher nutrient levels compared to conventional farming methods. Conventional farming practices, on the other hand, often rely on synthetic fertilizers and pesticides which can degrade soil quality and reduce the nutrient content of crops over time.

Soil erosion is another factor that can impact the nutrient content of food. Loss of topsoil through erosion can lead to depletion of essential nutrients, resulting in poorer quality produce. Sustainable agriculture practices, such as no-till farming and cover cropping, can help prevent soil erosion and maintain soil health for optimal nutrient content in cr. Soil microbiology also plays a crucial role in nutrient availability for plants. Beneficial soil microbes help break down organic matter and release nutrients for plant uptake, ultimately contributing to the nutrient content of food. Utilizing compost and other soil amendments can enhance the microbial activity in the soil, leading to improved nutrient levels in crops (Jones, 2016).

By understanding the relationship between soil quality, plant health, and nutrient content in food products, we can work towards implementing farming practices that prioritize soil health and sustainability to ensure nutrient-rich harvests for a healthy population. Soil testing methods can also be used to assess nutrient levels in food crops, informing farmers and policy makers on ways to improve soil quality and enhance the nutritional value of our food supply.

23.1 REGENERATIVE AGRICULTURE

Regenerative Agriculture (RA) is extensively utilized to describe agronomic methods based on the principles of enhancing soil health and sequestering carbon. Giller et al. (2021) give a comprehensive summary on RA from an agronomic perspective. Their review highlights that—ever since the RA concept entered regular practice in the 1980s—RA is a dynamic conceptual area which extends across agronomic, biophysical, and social justice dimensions. RA is usually framed in terms of agricultural systems which: (i) reduce the external effects of agriculture beyond the farm; (ii) reduce energy and other inputs into the farm; (iii) sequester carbon, increase nutrient cycling and wider ecosystem services, (iv) increase biodiversity, and (v) promote social justice. Soil conservation is thought of as the starting point for the majority of agronomic methods informed by RA, although universally accepted established definitions and inclusion criteria for RA are non-existent (Giller et al., 2021).

CHAPTER 24

SOIL DEGRADATION, A THREAT FOR FOOD SECURITY?

Contents

Food security is present when all people, at all times, have physical, social, financial, and economic access to enough, safe, and nutritious food that meets an individual's dietary requirements and food preferences for an active and healthy life. Nevertheless, the global population is increasing and is expected to reach approximately 10 billion by the year 2050, making the objective of food security for all even harder to reach and sustain. This level of population development would need to raise general food production by a minimum of 70% (Daszkiewicz, 2022; FAO, 2009).

24.1 SOIL DEGRADATION AND CLIMATE CHANGE

Soil degradation from climate change will have its greatest effect on food security. Soil degradation leads to a reduction in ecosystem services, lowers soil quality and also creates a loss of biodiversity which directly impacts human wellbeing. Erosion, salinization, pollution, loss of organic substance, excessive chemical and pesticide utilization, deforestation, urbanization and extreme mining activities are the cause for soil degradation. As an outcome of deforestation, the carbon/nitrogen equilibrium of the soil is deteriorated, plant nutrients and soil microbiological activity are reduced.

Curbing soil degradation is key to the food security and ecological maintenance of the Caribbean—yet unsustainable land management practices, linked with the pressures of climate change continue to affect soil health and threaten the regional prospects for future food security.

In accordance with a 2018 report from the United Nations Convention to Combat Desertification (UNCCD) 30% of degraded land in Small Island Developing States (SIDS) globally and more than 14% of the two billion hectares of degraded land worldwide are located in the Caribbean region.

The main reason for soil degradation in the arid, semi-arid, sub-humid and dry regions of the Latin America and Caribbean region is desertification, the process by which fertile land becomes desert, usually because of drought, deforestation, or inappropriate agriculture, which impacts 35% of the Caribbean region's surface area (Inter-American Institute for Cooperation on Agriculture (IICA), 2020).

Soil is at the root of food production and according to the Food and Agriculture Organization (FAO), soil degradation is one of the main reasons for reducing agricultural productivity.

The effect of human actions and climate change on soil health has direct implications on arable land, yields, and productivity of agricultural soils. The Caribbean is one of the most vulnerable regions in the world to climate change. Changes in rainfall patterns and temperatures, sea level rise, droughts, storms and flooding all have a function in the reduction of soil quality, especially that of the 20 centimeters of topsoil that is key for carbon sequestration, the process of capturing and storing atmospheric carbon dioxide.

Climate change effects, including hurricanes and droughts, increase land degradation and represent a major obstacle to the ongoing health of Caribbean soils and thus, food security.

Too much rainfall and wind experienced during a hurricane strips natural vegetation, leads to erosion and flooding and sets off landslides, flushing precious layers of topsoil out to sea.

For the 30-year period between 1991–2020, there was a 44% increase in said storms, a 20% increase in hurricanes and a 60% increase in major hurricanes in comparison to the previous 30 years. This augmentation in extreme weather has been linked to climate change.

During hurricane Maria in 2017, Dominica suffered approximately 25 kilo hectares of hurricane-related tree cover loss (63% of which was humid primary forest) translating into approximately 14 metric tons of CO_2 being released into the atmosphere. The agriculture sector was heavily affected, because of soil damage and a full loss of crops in addition to significant destruction of productive trees and the death of livestock in the region, droughts have additionally affected the health of soils, lowering its ability to sequester carbon, affecting agricultural yields and resulting in lower nutritional amounts in crops.

In 2014, Jamaica's most intense drought in three decades harmed farming communities and created a 30% drop in agricultural production. At the turn of the century, droughts affecting Guyana led to a 37% drop in rice production, and in Haiti, between 2015 and 2017, the food security of approximately 1 million people was made worse by on-going drought conditions.

As the Caribbean experiences fast development, these climate change effects have been compounded by unsustainable human actions.

Unsustainable agricultural practices, which cover tilling, deep root removal, slash and burn and indiscriminate application of harmful chemicals, have added to the degradation and depletion of 75% of natural carbon stocks and biodiversity from the area's agricultural soils (IICA, 2020).

49% of the land in the zone is exposed to water erosion and 56% is impacted by chemical degradation (salinity or acidity) of soils because of inappropriate agricultural methods.

In the case of Grenada, for instance, the excessive use of agricultural chemicals, including herbicides, pesticides and mineral fertilizers in cropland zones, has led to acidification across over 860 hectares of land (Grenada Land Degradation Neutrality Report, 2015).

Uncontrolled development and urbanization have added, almost ironically, to the degradation of soil and lowered food and water security. Lacking road construction, poorly managed settlements in landslide-prone areas and the growth of large residential and tourism-related structures have not only impacted the quality of soil and the availability of arable land, yet have also affected the quality of the area's water resources.

In Barbados for instance, between 2001 and 2020, the country missed out on 397 hectares or a 7.6% reduction in tree cover, the equivalent of 148 kilo tons of CO_2 emissions. A quarter of this tree cover loss took place in the parishes of St. James and St. Peter, located on what is called the Platinum Coast of the island, where the majority of the tourism activity takes place (Global Forest Watch, 2021).

At other locations in the region, mining has been an additional development-related reason for deforestation and desertification. In Haiti for example, 97% of all forests have been tilled and 4,200 hectares of soil have been eroded annually due to quick runoff—leading to the reduction of fertile topsoil and biodiversity loss.

Haiti's unsustainable land utilization practices have led to the irreversible erosion of a minimum of 6% of its land, with less than 20% of the land under cultivation considered to be suitable for agriculture (Global Environment Facility, 2018).

Similarly, in Jamaica, excessive mining of bauxite, the country's main mineral source, has led to soil erosion in the impacted region.

In accordance to research conducted by the Jamaica Environment Trust (2020), bauxitic soils hold considerable quantities of water, permitting the development of large trees and forests, yet when soils are removed for mining, soil fertility is gone, with effects appearing to be irreversible and not restorable by reclamation.

In accordance with estimates from the United Nations Environment Program in 2006, in the Small Island Developing States of the Latin America and Caribbean region, the degradation of land cost an approximate $4,800 million dollars per year, affecting about 125 million people.

From the point of view of food security, shallow and eroded soils are not sufficiently deep or healthy to support the area's indigenous crops. Root crops, legumes, and cash crops need soil that is more than 30 centimeters in depth. Dry pastures additionally lower the availability of feed for animals.

Box 1: A Solution-Based Framework for Hunger and Malnutrition in Nigeria

The Triple E framework covers problem identification found in the small-scale farmers of the rural communities of Nigeria and gives answers to the challenges. This framework is made up of three parts: Evaluation, Education, and Empowerment (Figure 24.1). The objective of the triple E is to champion rural agricultural activities, secure adequate and healthy food production, and strengthen food and nutrition security, especially among rural communities. This could be managed through the measurement and evaluation of the challenges linked with the production of food in a targeted group of people, capacity development through the education of farmers, and support for farmers with suitable tools.

The Triple E framework might be considered as a way to identify and solve challenges. It could help the government in pinpointing the true cause of the root problems causing minimal food production. All parts of the framework are key steps towards securing and achieving food and nutrition security. For Nigeria to be food secure, food production has to be suitable on a sustainable basis (Abdulwaliyu et al., 2023).

Any program targeted at augmenting food production in Nigeria must be assessed and backed up by a reduction in food inflation and dependence on imports, an increase in food exports, enhanced food and nutrition security, and a reduction in poverty within rural areas. In addition, the development team should be responsible for collecting and keeping statistics on the quantity of farmers, farms, and acreages, both arable and under cultivation.

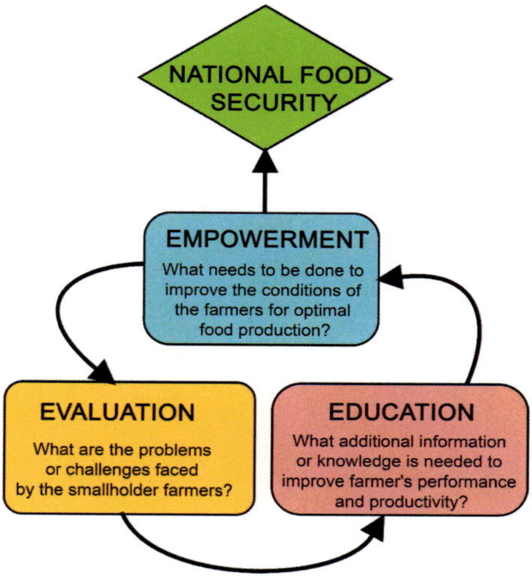

Figure 24.1. *Triple E framework for food security.*

Source: Abdulwaliyu et al. (2023).

CHAPTER 25

SOIL CONTAMINATION

Contents

Healthy soils are key for supporting human health. They are basic for the production of food, biomass and fiber, the production of some medicines, and maintaining and filtering water. Healthy soils also have a key function in carbon and nutrient cycles. Soil pollution impacts soil fertility; this puts food security at risk, which is key for human survival. It also presents risks to human health—indirectly via the intake of polluted food and drinking water, and directly via exposure to polluted soil.

Soils can turn contaminated in numerous manners. For instance, on-going pesticide application is an important issue since it reduces soil biodiversity and resilience, and could lead to the pollution of food and feed. The zero-pollution action plan sets targets of a 50% reduction in the utilization of chemical pesticides and the risks they pose, and a 50% reduction in the utilization of more dangerous pesticides by 2030. Excess nutrients are an additional source of soil contamination. They lead to ecosystem eutrophication and lowered biodiversity, and result from fertilizer and manure application or from the deposition of some air contamination. The zero pollution action plan establishes a goal to lower nutrient losses linked to too much nutrient application by a minimum of 50% by 2030 (for more information, see the zero pollution analysis on nutrient losses). Another goal which will promote healthier soils needs a reduction of 25% in the zone of land impacted by air contamination; this target is measured in the air and ecosystems section.

The EU soil strategy for 2030 reiterates the zero pollution objective that, by 2050, soil contamination should be so minimal that it no longer hurts human health (EC, 2021). The strategy prioritizes stopping contamination at its source, which aligns with the zero-pollution hierarchy. The soil strategy suggests a future soil health law; this is awaited to include provisions on identifying, maintaining an inventory of and remediating polluted locations—strongly lowering risks as a result.

As technological progress speeds up, electronic waste is thought of as a new emerging risk to soils. Fifty million tons of e-waste are created annually, turning it into one of the world's quickest growing contamination problems impacting our soils. FAO encourages communities to donate or recycle old devices instead of throwing them away.

While agrochemicals may aid in covering the global demand for food, they produce a legacy of soil contamination and lower quality agroecosystems. Approximately 60% of the top agricultural soils in 11 European countries have residues of various persistent pesticides. FAO encourages governments to promote the responsible and sustainable management of agrochemicals.

One third of the plastic produced worldwide ends up in the soils with plastic particles thus penetrating the food chain and the environment. Stronger environmental policies and regulations are required to avoid the utilization of too much packaging and chemicals by industry. Nevertheless, each and every individual may additionally add to resolving the challenge by just using a reusable bottle and grocery bag, and purchasing products with lowered or recycled packaging.

Healthy soil is key for human health. Healthy soil is paramount for the achievement of safe, healthy, and sufficient food. Healthy soil encourages strongly diverse ecosystems that give services critical to human survival, most importantly pollination. Healthy soil keeps water and protects waterways, in this way avoiding floods and waterborne diseases. Healthy soil maintains huge amounts of carbon and slows the speed of climate change.

Contamination of air, water, and soil is a significant and developing threat to global health. The *Lancet* Commission on Pollution and Health reported that contamination is the biggest environmental source of disease and premature death globally. Diseases emerging from contamination were the cause for approximately 9 million premature deaths in 2015—16% of all deaths globally and three times more than from AIDS, malaria, and tuberculosis together. In addition, contamination was responsible in 2015 for the loss of 268 million disability-adjusted life years (DALYs)—254 million years of life lost and 14 million years lived with disability.

Regardless of the fact that 70% of contamination-linked diseases are non-communicable diseases (NCDs) and over 60% are cardiovascular diseases, measures against contamination are minimally mentioned in the Global Action Plan for the Prevention and Control of NCD. As well as ignoring chemical contaminants, the Global NCD Action Plan also overlooks the health effects of non-chemical environmental factors including mental stress, noise exposure, nocturnal light pollution, and climatic changes that all together may easily outcompete all genetic predisposition-related health-risk factors. Epidemiological studies propose that environmental-risk factors are a main contributors to NCD, particularly cardiovascular disease, and that they also add to metabolic and mental conditions such as hypertension, heart failure, myocardial infarction, diabetes, arrhythmia, stroke, neurodegeneration, depression, and anxiety disorders in addition to cancers.

25.1. SOILS SUPPORT HUMAN HEALTH

Healthy soils allow for nutritious food, clean drinking water, raw materials and carbon sequestration functions—ecosystem services that are basic for ensuring food security, addressing climate change and securing human health (Montanarella and Panagos, 2021). Soil contamination impacts food security in two ways. To start with, soil contamination may lower crop outcomes; this is due to toxic contaminants breakdown soils over the long term. Next, soil contamination can turn foods harmful for human intake (Rodríguez-Eugenio et al., 2018).

Human exposure to soil contamination is considered to be partially responsible for over 500,000 premature deaths worldwide annually (Landrigan et al., 2018). The majority of these deaths happen in populations at risk, including children and the elderly, impacted by long-term exposure. Additionally, these deaths are linked to exposure to only a small number of pollutants; the effects of all soil contaminants on health and well-being are possibly larger.

Depending on the chemicals present, soil pollutants can impact numerous organs, including the lungs, skin, gut, liver and kidneys. These contaminants can also impact the immune, reproductive, nervous and cardiovascular systems, and more. Data has proposed that the health effects of soil contamination disproportionately impact lower income households; for instance, lower income households have greater opportunities of living near industrial sites and being exposed to polluted soils (Morrens et al., 2012; Levasseur et al., 2021).

Soil is key for human health in numerous forms. About 78% of the medium per capita calorie intake globally originates from crops grown directly in soil, and another nearly 20% originates from terrestrial food sources that depend indirectly on soil. Soil is also a main source of nutrients, and it performs as natural filters to remove pollution from water.

The lean crust of the Earth's surface backs up all terrestrial life and is part of the regulation and provision of numerous important ecosystem services that are basic to the environment and to human health and well-being. Soil is the bottom line of the agri-food system and the medium in which nearly all food-producing crops grow—about 95% of the food consumed originates from the soil. After the oceans, soil is the biggest active carbon store and one cubic meter of soil can keep up to 600 liters of water, allowing crops to grow over dry periods.

25.2. HUMAN EXPOSURE TO SOIL POLLUTION

When experiencing the outdoors, individuals can be exposed to contaminated soils via direct skin contact or by breathing in dust. Specifically, kids playing on the ground may eat or breathe polluted soil. Soils in urban locations, including parks and home gardens, are a fountain for pollutants (Li et al., 2018).

The amount of exposure is possibly due to weather and soil conditions, in addition to the distance to origins of contamination. These origins include agricultural fields on which pesticides, fertilizers, wastewaters and sewage sludge may be used; and industrial production, mining operation and waste management facilities, where dangerous materials may be utilized or kept—

making way for emissions to air, water and soil. People can additionally be indirectly vulnerable to soil contaminants via polluted drinking or bathing water. They could also eat polluted food grown in contaminated soils, or products from animals fed polluted feed. Additional information on contaminants in food is available in the chemicals and health testing.

Numerous of the new pollutants of concern in soils for example per- and polyfluoroalkyl substances (PFAS), phthalates and flame retardants are linked with health effects (Maddela et al., 2022).

Center points for human exposure to soil contamination are polluted locations, some agricultural and urban soils, and land that has been flooded before. A German investigation measured toxic pollutants in floodplain soils across the Elbe River in central Europe and evaluated risks for health (Rinklebe et al., 2019). The Elbe River and its tributaries are some of the most strongly contaminated rivers in Europe because of a history of forced industrial activity happening over time. Simultaneously, these floodplain soils are quite fertile and often utilized for agriculture. Around the Central Elbe River, arsenic was found to be the main source of total health hazard—followed by chromium, then vanadium and lead last. The study reported that kids' health was at noticeably higher risk than adults.'

Elevated and growing amounts of cadmium have been located in agricultural topsoil throughout Europe; this mostly comes from phosphate fertilizers. Cadmium is a heavy metal; food is the main exposure route in non-smokers while smokers are usually exposed to cadmium via cigarette smoke. It is connected to detrimental health impacts, including renal toxicity and osteoporosis (Ougier et al., 2021).

25.3. SOIL POLLUTION AROUND INDUSTRIAL INSTALLATIONS AND CONTAMINATED SITES

The majority of contaminants liberated into air or water from industrial emissions will end up in soils: a classical dispersion pathway is via surface run-off to streams and into soils, from which contaminants filter through to groundwater.

One recent investigation looked at the health effects on households living near three polluted mining and/or industrial locations in France, Portugal and Spain. The researchers reported evident impacts on health—such as lower birth weights, reduced health status among kids, elevated risk of chronic conditions among adults and higher premature mortality. Considering socio-economic costs, the study reported greater levels of school absenteeism and greater demand for health services among residents from contaminated locations, with lower income and less educated households being the most vulnerable to contamination (Levasseur et al., 2021). Second research looked at the health impacts of 1,544 waste landfill locations on local populations across Europe. It quantified that a total loss of 61,325 disability-adjusted life years (DALYs) per year was linked to health outcomes including low birthweight, congenital anomalies, respiratory diseases and annoyance from odor (Shaddick et al., 2018). A German investigation reported greater amounts of lead and cadmium in the blood of kids living in the vicinity of metal refineries in North Rhine Westphalia than in kids living elsewhere (Wilhelm et al., 2005).

CHAPTER 26

INTERCONNECTEDNESS OF SOIL HEALTH, PLANT HEALTH, AND FOOD NUTRIENT CONTENT

Contents

The interconnectedness of soil health, plant health, and food nutrient content is a crucial aspect of understanding the quality of our food supply. Soil quality plays a significant role in determining the nutrient content of the food that we consume. As food scientists, university students, researchers, and policy makers, it is essential to recognize the impact that soil health has on the nutritional value of our food.

One key factor in this relationship is the pH levels of the soil. Soil pH levels can have a significant impact on the availability of nutrients to plants, ultimately affecting the nutrient content of the food that they produce. Understanding the optimal pH levels for different crops can help to ensure that they are receiving the necessary nutrients to thrive and produce nutrient-rich food.

Organic farming practices have been shown to have a positive impact on soil quality, leading to higher levels of nutrients in food products. In contrast, conventional farming practices, which often rely on synthetic fertilizers and pesticides, can degrade soil quality over time, resulting in lower nutrient content in crops.

Box 1: The Planetary Health Diet

The planetary health diet is a worldwide reference eating pattern for adults that is illustrated by half a plate of fruits and vegetables. The complementary half is made up of mainly whole grains, plant proteins (beans, lentils, pulses, nuts), unsaturated plant oils, moderate quantities of meat and dairy, and some added sugars and starchy vegetables.

This eating plan is relatively flexible and permits adaptation to dietary requirements, individual preferences and cultural traditions. Vegetarian and vegan eating plans are two favorable options within the planetary health diet but remain personal choices.

The planetary health diet, also known as planetary diet or planetarian diet, is a flexitarian eating plan developed by the EAT-Lancet commission as part of a report released in *The Lancet* on 16 January 2019. The objective of the report and the eating plan developed from it is to establish dietary paradigms that have the following aims:

- To feed a world population of 10 billion people in 2050
- To significantly lower the global number of deaths due to poor diet
- To be environmentally sustainable as to prevent the collapse of the natural world

Soil erosion is another critical issue that can impact the nutrient content of food. Erosion can lead to the loss of topsoil, which is rich in nutrients, ultimately affecting the quality of crops grown in that soil. Implementing sustainable agriculture practices can help to prevent soil erosion and maintain soil quality for optimal nutrient content in food products.

Soil microbiology also plays a crucial role in nutrient availability for plants. Beneficial microbes in the soil help to break down organic matter and release nutrients that are essential for plant growth. Incorporating compost and other soil amendments can help to support healthy soil microbiology and improve nutrient content in food.

By understanding the relationship between soil health, plant health, and food nutrient content, we can work towards creating a more sustainable and nutrient-rich food supply for future generations. Implementing soil testing methods and sustainable agriculture practices can help to ensure that our food is not only delicious but also packed with essential nutrients for our health and well-being.

26.1. STRATEGIES FOR PROMOTING SOIL-PLANT HEALTH FOR NUTRIENT-RICH HARVESTS

In order to promote soil-plant health for nutrient-rich harvests, there are several key strategies that can be implemented. One important factor to consider is the soil pH levels, as they can have a significant impact on the nutrient content of food. Maintaining the proper pH levels through the use of organic farming practices can help ensure that plants have access to the necessary nutrients for optimal growth and development.

Organic farming practices have been shown to have a positive impact on soil quality, which in turn can lead to nutrient-rich harvests. By avoiding the use of synthetic pesticides and fertilizers, organic farmers are able to maintain a healthy balance of nutrients in the soil, resulting in healthier plants and higher nutrient content in the food they produce.

In addition to organic farming practices, the use of compost and other soil amendments can also help improve nutrient content in food. Compost is rich in organic matter and essential nutrients, which can help replenish the soil and provide plants with the necessary resources for healthy growth. By incorporating compost into the soil, farmers can improve soil quality and ultimately increase the nutrient content of their crops.

Sustainable agriculture practices are also essential for maintaining soil quality and nutrient content in food. By implementing practices such as crop rotation, cover cropping, and reduced tillage, farmers can help prevent soil erosion and maintain a healthy balance of nutrients in the soil. These practices can help ensure that plants have access to the necessary nutrients for optimal growth and development, resulting in nutrient-rich harvests.

Overall, promoting soil-plant health for nutrient-rich harvests requires a holistic approach that takes into account the relationship between soil health, plant health, and nutrient content in food products. By implementing strategies such as organic farming practices, the use of compost and other soil amendments, and sustainable agriculture practices, farmers can help ensure that they are producing nutrient-rich food that supports the health and well-being of consumers.

26.2. IMPLICATIONS FOR POLICY AND RESEARCH IN ENHANCING FOOD NUTRIENT QUALITY

As we delve deeper into the relationship between soil quality and food nutrient content, it becomes evident that there are significant implications for policy and research in enhancing the nutrient quality of our food supply. This has important implications for food scientists, university students, researchers, and policy makers alike.

One key area for policy intervention is the promotion of sustainable agriculture practices that prioritize soil health. This includes implementing measures to reduce soil erosion, such as contour plowing and cover cropping, which can help maintain nutrient levels in the soil and ultimately in the crops grown in it. Policy makers can also incentivize the use of compost and other soil amendments to improve soil fertility and nutrient availability for plants.

Research in this area can help to further elucidate the impact of soil pH levels on food nutrient content, as well as the differences in nutrient content between organic and conventional farming practices. Understanding the role of soil microbiology in nutrient availability for plants is also crucial for developing strategies to enhance food nutrient quality.

Furthermore, research into soil testing methods can help to assess nutrient levels in food crops more accurately, leading to better-informed policy decisions regarding food production and distribution. By studying the relationship between soil health, plant health, and nutrient content in food products, researchers can identify ways to optimize nutrient levels in our food supply.

Overall, by focusing on enhancing soil quality and nutrient content in food, we can work towards a more sustainable and nutrient-rich food system. This requires collaboration between researchers, policy makers, and practitioners in the fields of soil science, agriculture, and nutrition to ensure the health and well-being of both people and the planet.

CHAPTER 27

IMPROVING FOOD PRODUCTION AND NUTRITION SECURITY

Contents

Empowerment in agriculture plays a key function in improving food production and nutrition security (Mwololo et al., 2022). Thus, farmers in rural communities might require empowerment or be provided with the following:

- **Suitable Security:** suitable security is a required tool for suitable food production, and no country can be food secure in the middle of insecurity, particularly in rural communities (Abdulwaliyu et al., 2023).
- **Non-Corrupt Policies and Programs that Will Turn Farming More Profitable:** suitable policies are key components for food sovereignty and can significantly impact consumer demands (Sabaté et al., 2016). One research reported that a single percent augment in farmers' access to agricultural finance is connected with an increase in food production by 0.002 to 0.006%; more agro-funding at minimal rates would also incentivize the farmers (Osabohien et al., 2020). Regardless of the recognized effect of agro-financing on the production of food, the government must lower or desist from providing loans to non-farmers. Non-farmers are likely to benefit more from monies or loans meant for farmers, thus desired results are not often met. If the support for farmers were material-based, such as mini-mechanical farm equipment, seed, etc; most of this problem could be avoided. No matter how favorable a policy is and the quantity of money involved, it will be useless amidst the NO-STEP (no sincerity, transparency in execution of policy) syndrome. This syndrome is the main cause, or perhaps the only cause, why numerous countries are developing at a slow pace. If the government does not get better, calls for development in the agricultural sector would add up to not much, and food security would most likely not be achieved.
- **Higher-Yielding Crop Varieties:** it has been shown that a higher crop yield strongly reduces hunger rates (Ngoune and Shelton, 2020).

The distribution or subsidy of food have always been huge strategies to improve the welfare of food-insecure communities (Shittu et al., 2018), particularly for people from rural areas, yet sustainability is always a problem. Thus, giving the food-insecure household the required tools, knowledge, and skills to empower them to produce suitable food on a sustainable basis is a better option.

The COVID-19 pandemic and its economic impact continues to affect international food systems with negative effects on food security and nutrition, especially in low- and middle-income countries. It is predicted that the number of people in acute food insecurity will most likely have doubled between 2019 and 2020. Restrictions at borders and lockdowns put planting, harvests and processing at risk, limit transport to markets, reduce incomes and alter food services and retail.

For farmers not limited by labor shortages, subsidizing inputs may champion the utilization of fertilizer, modified seeds, pesticides, fuel and equipment. Because shortages of labor force can turn into a recurring problem, extra labor supply interventions including "green corridors" for migrant workers (European Commission, 2020) or, more permanently, increases in mechanization (Swinnen and Mcdermott, 2020) can be required.

Because of limitations in manpower, market closures and modifications in downstream procedures and retail, whole harvests may be lost before getting to the consumer. Post-harvest, storage and processing interventions can reduce such losses in terms of both quality and amounts. Temperature-controlled supply chain technologies can strongly increase the shelf life of vegetables (Kumar et al., 2004), or turn supply chains for dairy more resilient (Trotter and Mugisha, 2020).

The COVID-19 emergency highlighted pre-existing problems in the food system. Investment in infrastructure for transportation, power, irrigation and storage facilities is required to keep its activities. Financial and technical backup for small and medium-sized enterprises (SMEs), covering the setup and maintenance of e-commerce, is especially important to support consumers' food and nutrition security (FAO, 2020a).

27.1. EFFECTS ON FOOD AND NUTRITION SECURITY

Information interventions as well as input subsidies augment the productivity, yields and income of farmers (Takahashi, Muraoka and Otsuka, 2019; Hemming et al., 2018), and can thus augment food availability during large scale disruptions. Nevertheless, evidence for direct beneficial impacts on food security and nutrition is inconclusive (Pace et al., 2018; Harou, 2018; Walls et al., 2018). Lower income households benefit disproportionately in regards to production increases from combining input subsidies with additional social transfers (Pace et al., 2018). Nevertheless, the long-term effectiveness of input subsidies relies on the prevention of crowding-out of demand in commercial markets (Jayne et al., 2018).

Agricultural practices that are also nutrition-sensitive may improve numerous health outcomes in mothers and children. For example, both explaining the importance of nutritious eating practices in extension services while giving biofortified seeds (Ogutu et al., 2020) augment use in addition to overall nutritional status. Research reviews have reported that home gardening and, with less evidence, urban gardening interventions have a beneficial impact on the availability of nutritious foods and dietary diversity (Galhena et al., 2013; Poulsen et al., 2015).

Where production diversity is minimal, access to markets is a prime factor in augmenting dietary diversity (Ruel et al., 2018). Investment in infrastructure may strengthen supply chains, lower transportation costs and increase market access, thus increasing food availability, access and stability. In accordance with a comprehensive review, investments in infrastructure may lead to even larger reductions in poverty than input subsidies (Jayne et al., 2018). Better post-harvest handling, storage and processing practices lower food losses and strengthen food security via an increase in food availability and stability (Kumar and Kalita, 2017).

School meals given in the form of take-home rations or in-kind food transfers may increase the availability of and access to food for schoolchildren and their families (Ruel et al., 2013). In-kind food transfers have been reported to augment the caloric intake of the recipient household. Cash transfers augment both caloric intake as well as dietary diversity (Durao et al., 2020; Gentilini, 2016; Hidrobo et al., 2018) directly by increasing access to food and indirectly via improved nutrition decisions (Burchi et al., 2018). In the long term, cash transfers and vouchers are considered to be more efficient in improving overall nutrition status (Gentilini, 2016).

CHAPTER 28

COMPARISON OF "ENVIRONMENTAL IMPACT, NUTRITIONAL QUALITY, AND PRICE AMONG SIX POPULAR DIETS

Contents

Sustainable eating practices and their general effect on the environment and human health have been gaining international attention. With the augmenting demands of an increasing worldwide population and the impact of climate change, personal as well as collective dietary selection is paramount to a viable future (Jarmul et al., 2019). In accordance with the Food and Agriculture Organization (FAO), sustainable eating practices are defined as nutritionally adequate, safe, healthy, economically affordable, culturally acceptable, and imposing minimal environmental degradation on Earth (Food and Agriculture Organization of the United Nations, 2023). Sustainable eating practices are, thus, a mix of food selection that guarantee the protection of human health, human rights, as well as the environment.

The term "sustainability" makes reference to three measurable indicators of a sustainable eating practice as established by the FAO—environmental impact, nutritional quality, and affordability. To begin with, environmental impact makes reference to ecological modifications brought by food production, harvest, transport, and storage. Next, nutritional quality makes reference to the relative presence of key nutrients as well as to unfavorable components in a food item. The nutritional value of commonly consumed food items is key to consider as it adds to the physical health and development of human populations. Finally, affordability of food items makes reference to the accessibility of eating practices and their subsequent implementation. Consideration of these three metrics within the frame of sustainability aids in understanding the short and long-term impacts of food system methodologies on both human consumption and environmental health.

In a 2014 study, Masset et al. assessed the sustainability of foods commonly consumed by the population of France utilizing these specific three dimensions of sustainability. Each food item was evaluated for sustainability by calculating a score based on environmental impact indicators, nutritional quality, and price. To evaluate the environmental impact of foods, the researchers utilized three environmental indicators: greenhouse gas emissions (GHG), air acidification, and freshwater eutrophication. They reported that meat, fish, eggs, and dairy had the greatest impact on the environment (Masset et al., 2014). In addition, foods with greater environmental impact, especially GHG emissions, were also usually the least affordable and had reduced nutritional quality per kilogram. The majority of plant-based food items had greater overall sustainability scores when compared to animal-based products (Masset et al., 2014).

Certain data has demonstrated that the Mediterranean diet supports an increased life expectancy due to its high nutritional quality (Sezaki et al., 2022). The ketogenic diet is considered to improve chronic conditions by removing sugars and carbs and leading to a state of ketosis in the body; in an effort to take up a healthier lifestyle and reduce fat, some people select the ketogenic diet

(Appavu et al., 2016). Vegan and vegetarian eating plans, which are made up of mostly plant-based products, are thought to be among the best for environmental sustainability (Chai et al., 2019; Segovia-Siapco and Sabaté, 2019; Rabès et al., 2020). Although some vegans and vegetarians make food selections based on perceived environmental sustainability, some additionally value animal welfare, another key factor of food system adequacy (Hopwood et al., 2020). The paleo eating plan promotes a return to the "natural" state of human eating practices prior to the Neolithic revolution by limiting intake of all processed foods. Some consider this to be the healthiest state for humans because of the evolution of the metabolic systems (Harvard T. H. Chan School of Public Health, 2023). An eating plan that follows WHO nutritional guidelines is provided for the reader as a general guideline for a healthy eating plan based on the best available nutritional information to date (World Health Organization, 2018).

28.1. PROPORTION OF FOODS OF ANIMAL ORIGIN

Over half of the food (grams) eaten in a week of the ketogenic diet is made up of foods of animal origin, while the WHO eating plan nearly reached the 50% mark. For each of the other four eating plans, only approximately 20–30% of their food intake in grams originated from foods of animal origin, showing lower meat product consumption in comparison to the ketogenic and WHO eating plans. Of note, the Mediterranean and Paleo diets had a larger proportion of kilocalories obtained from meat-based products in comparison to the vegan and vegetarian diets, both reaching the 30–40% levels and hence approaching the WHO diet's ~45% of kilocalories derived from foods of animal origin.

It is worth mentioning that the vegetarian eating plan, even with the removal of meat, included a strong amount of foods of animal origin due to the inclusion of eggs, butter, honey, and cheese. Other eating plans included both meat and meat derived products. Vegan and vegetarian eating plans stand as the eating plans with the complete removal and lowest amounts of, respectively, foods of animal origin, while the Mediterranean and Paleo eating plans have moderate quantities of meat and animal derived products. The WHO and ketogenic eating plans included the greatest quantities of meat product in grams, with about half of the ketogenic diet made up of foods of animal origin in both grams and kilocalories.

28.2. ENVIRONMENTAL SUSTAINABILITY

Eating plans containing meat had the strongest environmental impact, while plant-based eating plans had the least. The ketogenic eating plan had the greatest total GHG emissions and largest emissions per gram of food, followed closely by the WHO eating plan. Vegetarian and vegan eating plans resulted in the lowest GHG emissions per gram of food, with nearly half or less GHG emissions in comparison to the meat-heavy ketogenic eating plan. The Paleo eating plan also had high GHG emissions, though not as much as ketogenic and WHO eating plans.

The ketogenic eating plan resulted in the greatest total land utilization per gram of food, followed by the WHO eating plan, while the vegan eating plan resulted in the lowest land utilization. Outside the WHO eating plan, all diets just reached half the land utilization of the ketogenic eating plan. Vegetarian and Mediterranean diets had comparable land use per gram, while the Paleo diet came closer to reaching the average line for total and mean values.

The ketogenic eating plan contributed strongest to water eutrophication, closely followed by the WHO and paleo eating plans, respectively, while the vegan eating plan contributed the least to eutrophication. Even though the vegetarian eating plan also contributed toward water eutrophication, closely following the Mediterranean diet, it still had a lower impact when compared to the WHO, ketogenic, and paleo eating plans. On note, the paleo eating plan's total eutrophication potential reached beyond the overall total average, showing that it will not obtain a sustainability point, but the mean eutrophication potential remains lower than the overall mean average; thus the distinct relative sustainability scores when comparing total values and mean values.

For water withdrawals, the WHO and paleo eating plans, correspondingly, had the greatest total water utilization among all eating plans. Per gram, the ketogenic diet had greater water withdrawals, followed by the paleo and WHO eating plans. Vegan and vegetarian diets had comparable total and mean water utilization, the lowest among the eating plans, while the Mediterranean eating plan stood at a mid-point between low and high utilization.

Generally, the meat-dense diets (ketogenic diet and WHO diet) had greater values for the four environmental impact indicators, while those characterized by less meat products or elimination of meat exhibited mid-level values for these indicators. The vegan eating plan had the lowest values across all environmental indicators.

28.3. NUTRITIONAL QUALITY

The paleo eating plan had the greatest total nutritional quality across all eating plans, and the vegan and Mediterranean eating plans followed closely; there were no significant patterns among the three. Eating plans that had a diversity of food types and reduced or eliminated the intake of meat products (vegan, paleo diet, Mediterranean diet) had the greatest nutritional quality. The ketogenic eating plan almost surpassed the general average, performing better than the previous metrics.

Notably, the eating plan rankings change when nutritional quality per gram is explored. Diets that limit or remove the consumption of entire food groups (such as meat elimination in the vegan eating plan and carbohydrate restriction in the ketogenic eating plan) appear to be more nutritious per gram. Vegetarian and WHO eating plans followed behind; the ketogenic eating plan, however, surpassed the overall mean average and obtained its only point in the relative sustainability score calculated with mean values.

28.4. AFFORDABILITY

The WHO eating plan had the greatest overall cost among all eating plans, while the vegetarian eating plan was the least expensive both in total cost and in cost per gram. Per gram, the ketogenic eating plan was the most pricy, closely followed by the paleo eating plan. While the ketogenic eating plan scored a sustainability point for total cost as it stayed relatively affordable in comparison to the other eating plans, it did not score a point for its price per gram. The Mediterranean eating plan scored a point for having a price per gram lower than the overall mean average but did not score a point for having a high overall price.

Plant-based eating plans (vegan and vegetarian) and plant-forward eating plans, which contain less animal products (Mediterranean) have greater relative sustainability when in comparison to eating plans that include moderate or high consumption of foods of animal origin, a finding supported by previous research (Tilman and Clark, 2014; Lacour et al., 2018; Segovia-Siapco and Sabaté, 2019). Previous studies have shown that following a plant-based eating plan usually results in a smaller environmental footprint (Lacour et al., 2018). The heavily meat-based ketogenic eating plan and the WHO eating plan received the lowest sustainability scores across environmental impact, nutritional quality, and affordability metrics.

In climate studies, GHG emissions regularly provide a strong indicator of environmental impact as these emission levels are directly linked with other environmental impact indicators, including air acidification and freshwater eutrophication, hence affecting the environment in its entirety (Röös et al., 2013). The higher GHG emissions of the ketogenic eating plan when compared to the vegan and vegetarian eating plans are consistent with previous research that report the negative effect of meat production on the environment (Tichenor et al., 2017; González et al., 2020; Grosso et al., 2020). Thus, results demonstrating higher GHG emissions for the meat-heavy ketogenic eating plans and the positive correlation between GHG emissions and the proportion of foods of animal origin in an eating plan was expected.

The eating plan based on WHO guidelines resulted in the second greatest environmental footprint following the ketogenic eating plan. This finding was not entirely surprising since the WHO eating plan also involved regular intake of foods of animal origin. For both the WHO eating plan and the ketogenic eating plan, foods of animal origin represented an elevated amount of food consumed over a single week period both in grams and in kilocalories. The elevated GHG emission and eutrophication amounts of the WHO and ketogenic eating plans are consistent with previous studies demonstrating that meat products and processed foods have strong impacts on freshwater eutrophication because of runoff and waterway contamination (Tusseau-Vuillemin, 2001; Xue and Landis, 2010). In addition, the paleo eating plan scored a sustainability point for mean eutrophication potential, but not for total eutrophication potential. This suggests that the paleo eating plan contains some food items that add significantly to eutrophication development, but overall, they are offset by other food items that make up most of the meal plan and do not add to eutrophication potential. A person following the paleo eating plan might find it simple enough to reduce their eutrophication footprint by just avoiding higher-contributing items.

Likewise, the discovery that ketogenic and WHO eating plans have the greatest land utilization is in accordance with studies reporting that the production of meat requires greater physical space than other agricultural practices, particularly when considering feed for livestock (Hallström et al., 2014; Alexandre et al., 2017). even though land utilization is strongly and directly linked with meat intake, the presence of foods of animal origin does not seem to be a predictor of water withdrawals, a finding backed up by investigations demonstrating that water productivity has improved in the livestock sector (Mekonnen et al., 2019) and that certain food items including nuts, fruits, and vegetables need greater amounts of water when compared to livestock (Aleksandrowicz et al., 2016). Eating plans with greater amounts of foods of animal origin (ketogenic, WHO, and paleo eating plans) did not score a sustainability point for water withdrawals since their total and mean values were much greater than the other eating plans. The vegan and vegetarian eating plans, for instance, utilize about half of the total water needed by the WHO and paleo eating plans.

Intake of red and processed meats add to the risk of cardiovascular disease and all-cause mortality (Wolk, 2017); certain characteristics of red and processed meats, including saturated fat and sodium, are considered as unfavorable nutrients. Thus, the removal of processed meats in the paleo eating plan and the strong limitation of red meat in the Mediterranean eating plan, coupled with the removal of both in the vegan and vegetarian eating plans, could minimally partially explain the reasons behind these eating plans having greater nutritional quality in comparison to an eating plan based on the WHO guidelines. Foods of animal origin that are not meat, including dairy, do not demonstrate a similar correlation with all-cause mortality and are not linked with cardiovascular disease (Farvid et al., 2017).

The ketogenic eating plan had interesting results. For example, some food items found in this eating plan have a rich nutritional profile, yet they are not eaten in sufficient quantities and are offset by the consumption of items that have lower nutrient profiles. This is in accordance with the ketogenic eating plan's characterization of elevated intake of lean meats including fish and chicken and non-starchy vegetables, but in addition to items elevated in saturated fat including bacon, ham, and similar. Additional investigation on the nutritional quality and physiological effects of red meat, processed meats, and meat intake overall may give insight into whether meat and meat products are required to meet nutritional needs.

While price was not linked with the quantity of foods of animal origin, the total and mean results for each eating plan may reflect the elevated cost of meat when in comparison to grains, legumes, and other plant-based, energy-dense foods. Vegan and vegetarian eating plans had the least total price and price per gram when in comparison to the other eating plans, but minimally. No conclusions can be inferred because of a lack of statistical significance. In terms of total cost, the WHO, paleo, and Mediterranean eating plans were, correspondingly, the most expensive, with paleo and WHO also not scoring a point for mean cost.

The ketogenic eating plan scored a point for its overall cost, which is below the overall total average, likely due to its high relative proportion of calorie-dense products. This means lower amounts of food, in grams, is required to meet 2,000 kilocalories/day. Nevertheless, the cost per gram of the ketogenic eating plan is the greatest among all eating plans, which means the specific food items, overall, are relatively expensive when compared to the specific food items of other eating plans. The Mediterranean eating plan demonstrated the opposite pattern: total cost was elevated and did not earn the diet a sustainability point, but mean cost was lower than the overall mean average. This means that certain food items found in minimal amounts of the Mediterranean eating plan are constantly enough to increase the total price, but on average, the more frequently consumed items are not as expensive.

These inconclusive outcomes may be explained by the elevated price of fresh fruits and vegetables, whole grains, lean meat, and other food items thought of as "healthy," which are generously found across all eating plans with the exception for the eating plans that remove meat (Drewnowski, 2010; Darmon and Drewnowski, 2015). Removing a costly food group may have positive effects on the plant-based eating plan's affordability, even if the eating plan still covers other expensive food groups. Furthermore, the relative affordability of various diets may vary by region. In some locations, foods of animal origin are less costly when compared to fresh produce, highlighting the requirement for region-specific analyses.

The WHO's guidelines are aimed at making dietary recommendations based on the nutritional value of food items. While nutrition is a key consideration, the organization has missed to suitably consider affordability and environmental impact. It might be favorable to add environmental impact and affordability along with nutritional quality when designing dietary guidelines and recommendations. The WHO and other organizations can also consider providing more comprehensive recommendations by adjusting them to specific regions. While the nutritional quality of eating plans must stay consistent, the environmental impact measures and financial costs will likely fluctuate by region.

While four different environmental sustainability indicators were considered, scientists aiming to expand on the topic may give a more robust analysis of environmental sustainability by covering extra indicators, for example air acidification. Another option to expand the research could be to look at variations of the same eating plans included here. Diets such as the Mediterranean, vegetarian, or WHO diets are relatively flexible, and variations of a week's worth of food could significantly differ from what was recommended by the meal generator, hence yielding different results. Furthermore, because this work prioritized the overall meat consumption within each eating plan, no distinction was made between red meat, poultry, dairy, and other foods of animal origin when calculating the proportions and correlations of the various eating plans. A next step could be to revise if there are more significant correlations according to the category of animal product, as red meat has different cost and nutritional values from poultry, fish, and other types of meat.

As innovative farm-to-fork food solutions become more common and taken up, it might be useful to revisit these analyses, considering new agricultural practices, food production methods, packaging, and technologies for shipping. As an example, the production of meat from stem cells is considered much more efficient and environmentally sustainable, reducing the amount of water needed, land, and greenhouse gas metrics in comparison to traditional meat production (Hadi and Brightwell, 2021).

The standard American diet, more often referred to as the "Western Diet" in the literature, is not so much an intentional eating plan and is most often centered on food for sustenance rather than prioritizing health (Rakhra et al., 2020). The standard American diet is discussed here as a control for comparison to the other eating plans. The Western Diet became popular as a meal plan (and has continuously grown) since the early 1990s. Strong economic development and the internationalization of western supply chains for food allowed for the expansion of modern industrial manufacturing practices in the food industry, leading to a new standard for access and distribution (Azzam et al., 2020). Several of these foods covered artificial sweeteners, engineered oils, and syrups that would come to rule the available foods in the market, turning them into staples in what is now referred to as the diet of the standard American. Not many would claim that they adhere to the "Standard American diet," thus the number of consumers of this eating plan must be estimated. The Center for Disease Control and Prevention (CDC) stated that "in 2015–2018, 17.1% of U.S. adults aged 20 and over were on a special diet on a given day" (CDC, 2022). With this information in mind, and for the purposes of this discussion, it can be assumed that the large majority of the U.S. population (above 80%) are not following specific eating plans and follow what could be referred to as the standard American diet.

The eating practices of a lot of Americans are regularly described as poor in suitable and sufficient portions of fruits, vegetables, and whole grains in addition to excessive quantities of ultra-processed foods and a significant intake of salt, sugar, and saturated fats (CDC, 2022). Hence, the eating plan is inherently lacking in several of the valuable micronutrients, macronutrients, vitamins, and minerals required for top health. Additionally, western food traditions often look for "highly palatable foods that can trigger eating addictive-like behaviors" (which are usually satisfied by ultra-processed or pre-packaged food items) and lead to numerous preventable diseases including "atherosclerosis, type II diabetes, hypertension, and, obesity" (La et al., 2018; Carpenter, 2013). The well-known and ever more common obesity epidemic within the U.S.—In 2018 the obesity prevalence was reported at 41.9% in the U.S. population of adults—is evidence of the harm that may be originating from a calorically rich while nutrient-poor eating pattern (Chiba, 2019). The high and frequent consumption of ultra-processed food products linked with low quantities of whole foods coincides with a plethora of negative health effects including heart disease, metabolic and gastrointestinal disturbances, neurodegenerative diseases, diabetes, and infertility, amongst others (Jayedi, 2020). Additionally, since the Western Diet is often considered to be highly unhealthy, those that consume the American eating plan regularly are more likely to lack regular physical activity and are more likely to consume alcohol and drugs, which lead to health conditions including coronary heart disease, diabetes, obesity, and depression (Matison et al., 2021; USDA, 2022).

CHAPTER 29

POLICY IMPLICATIONS FOR PROMOTING SUSTAINABLE FARMING PRACTICES

Contents

In order to promote sustainable farming practices that enhance soil quality and food nutrient content, it is essential for policymakers to prioritize the implementation of supportive policies. These policies should aim to incentivize and support farmers in adopting practices that prioritize soil health and nutrient-rich harvests.

One key policy implication is the need for government subsidies and financial incentives for farmers who implement sustainable farming practices. This could include subsidies for the use of organic farming methods, which have been shown to improve soil quality and increase nutrient content in food crops. By providing financial support to farmers who prioritize soil health, policymakers can encourage the widespread adoption of sustainable practices that benefit both the environment and human health.

29.1 THE ROLE OF POLICY MAKERS

Additionally, policymakers should consider implementing regulations that restrict the use of harmful chemicals and pesticides in agriculture. These substances can degrade soil quality, reduce nutrient content in food crops, and harm human health. By enforcing regulations that promote the use of natural and organic farming methods, policymakers can protect soil health and ensure the production of nutrient-rich food for consumers.

Furthermore, policymakers should invest in research and education programs that provide farmers with the knowledge and resources needed to implement sustainable practices. By funding research on soil microbiology, composting, and other soil improvement techniques, policymakers can help farmers make informed decisions about how to enhance soil quality and nutrient content in their crops.

Overall, promoting sustainable farming practices through supportive policies is essential for improving soil quality and food nutrient content. By prioritizing soil health and incentivizing sustainable practices, policymakers can ensure the production of nutrient-rich harvests that benefit both the environment and human health.

CHAPTER 30

SMALL FARMERS AND THEIR CONTRIBUTION TO THE ALLEVIATION OF FOOD INSECURITY

Contents

The Food and Agricultural Organization and the World Health Organization have highlighted that small scale farmers are the key frontline players for reaching sustainable food and nutrition security, and empowerment of the rural agriculture system as the main pathway towards sustainable food security (Dioula et al., 2013). Thus, providing the required support to small scale farmers could increase their production capacities, which may have beneficial effects on nutrition and food security at local, regional, and national levels (Louhichi et al., 2020). If rural farmers remain on the sidelines, it is likely that in the near future governments may have no option but to subsidize food. The scenario is likely to play out as follows: only those at the high end of the social ladder would have the capacity to import foods or produce locally, thus would sell them for any price. Governments must learn from the lessons of the West, where foods are strongly subsidized. A study by Nordin and Nordin (2017) reported that, in the majority of the world, governments are currently spending billions of dollars on food subsidies. And since numerous foods are poor in certain nutrients, governments are also pushed to spend billions of dollars to subsidize nutritional management through fortification, supplementation, and medical interventions (Nordin and Nordin, 2017).

Smallholders control more than 80% of the global estimated 500 million small farms and grow more than 80% of the food eaten in a significant section of the developing world, adding strongly to reducing poverty and food security. Augmenting fragmentation of landholdings, in addition to lower investment support and marginalization of small farms in economic and development policy, risks this contribution, increasing the vulnerability of smallholders.

Smallholder farming practices, in turn, affect the condition of ecosystems. These impacts are not always negative, but poverty and immediate needs can drive smallholders to put pressure on ecosystems, for example through habitat modification, over extraction of water and nutrients, and use of pesticides.

30.1 INTERNATIONAL DEMAND FOR FOOD

The demand on agriculture to feed a greater and more urbanized population by way of international markets in the coming 40 years will keep evolving, placing extra pressure on available land and additional natural resources. Modern practices are compromising the ecological foundation of the international food system via overuse and the impacts of agricultural pollution, thus increasing degradation, lowering ecosystem capacity to create sustainable outputs and threatening to affect food security and poverty reduction in an unfavorable manner. Sustainable agricultural intensification may be the solution to increase food security, protect the environment and reduce poverty.

With their significant collective experience and detailed knowledge of local conditions, smallholders have numerous practical solutions that can aid in placing agriculture on a more sustainable and balanced footing. For this to happen, they require help to face market challenges and other disincentives for sustainable land utilization, including insecure land tenure, elevated transaction costs and weak institutional support. A key challenge will be to face the differences of scale between decisions made at the farm level and effects at larger ecosystem scales.

Additional studies are required for the following:

- **Biodiversity/Ecosystem Services:** Greater amounts of information are required regarding the relationship and dynamics across biological communities and the services they offer, including how these relationships develop over time and how this impacts the stability and resilience of the services and agricultural productivity.
- **Synergistic Effects of Below- and Above- Ground Services:** Whether suites of below- and above-ground services add synergistically, or trade off in their contribution to crop yield and quality remains unknown.
- **Impacts of Land Use on Biological Communities:** For optimal handling of multiple services, greater data is required on how land utilization and other environmental factors impact the distribution, abundance and community structure of organisms that support crop production.
- **Climate Change and Agricultural Productivity:** Comprehension must be deepened on the effects of climate change on agricultural output, cropping practices, the spread of crop disease, disease resistance, and irrigation development.
- **Economics of Sustainable Intensification:** So that ecosystem services become an integral part of farming, further information is required into the economic perks and costs connected with ecological intensification.

CHAPTER 31

FUTURE DIRECTIONS IN UNDERSTANDING THE LINK BETWEEN SOIL QUALITY AND FOOD NUTRIENT CONTENT

The link between soil quality and food nutrient content is a complex and multifaceted relationship that requires further exploration and research in order to fully understand its implications. As we have discussed throughout this book, soil quality plays a crucial role in determining the nutrient content of the crops that we consume on a daily basis.

Moving forward, it is important for food scientists, university students, researchers, and policy makers to continue to investigate the various factors that influence soil quality and its impact on food nutrient content. This includes studying the effects of soil pH levels, organic versus conventional farming practices, soil erosion, soil microbiology, and the use of compost and other soil amendments on the nutrient content of food.

Furthermore, sustainable agriculture practices must be implemented in order to maintain soil quality and nutrient content for future generations. This includes using soil testing methods to assess nutrient levels in food crops, as well as understanding the relationship between soil health, plant health, and nutrient content in food products.

Overall, the future direction in understanding the link between soil quality and food nutrient content lies in continued research, collaboration, and education within the fields of soil science, agriculture, and nutrition. By working together to address these important issues, we can ensure that the food we eat is not only delicious, but also nutrient-rich and beneficial for our health and well-being.

CHAPTER 32

THE ROLE OF AI ON FOOD PRODUCTION

Contents

Artificial Intelligence (AI) has revolutionized the field of food production in recent years, offering innovative solutions to address various challenges faced by the agriculture industry. From precision agriculture to food processing and packaging, AI technologies have significantly enhanced the efficiency, productivity, and sustainability of food production systems. This subchapter explores the profound impact of AI on food production and its implications for nutritionists, farmers, food scientists, and food technologists.

AI and its role in food production have transformed traditional farming practices by enabling farmers to make data-driven decisions in real-time. By leveraging AI applications in precision agriculture, farmers can optimize crop yields, reduce resource wastage, and minimize environmental impact. These AI-driven solutions not only enhance productivity but also contribute to sustainable farming practices, ensuring the long-term viability of agricultural systems.

In the realm of food processing and packaging, AI-driven automation has revolutionized the way food products are manufactured, packaged, and distributed. By implementing AI technologies, food companies can streamline production processes, improve product quality, and ensure food safety. These advancements not only benefit consumers by providing them with high-quality, safe, and nutritious food products but also enhance the overall efficiency of the food supply chain.

Moreover, AI applications in sustainable farming practices have enabled farmers to adopt more environmentally friendly and resource-efficient farming methods. By utilizing AI- powered tools and technologies, farmers can monitor soil health, optimize water usage, and reduce chemical inputs, leading to improved crop yields and reduced environmental impact. These sustainable farming practices not only contribute to the preservation of natural resources but also promote the production of healthier and more nutritious food.

Furthermore, AI-driven personalized nutrition and dietary recommendations have revolutionized the way individuals approach their dietary habits and food choices. By analyzing individual health data, dietary preferences, and nutritional requirements, AI technologies can provide personalized recommendations for optimal nutrition and diet planning. These personalized nutrition solutions not only empower individuals to make informed choices about their food intake but also promote a healthier and more sustainable food consumption pattern.

In the realm of food waste reduction and management, AI-powered solutions have significantly improved the efficiency of food waste management processes. By implementing AI technologies, food companies can track, analyze, and optimize their food production and distribution processes to minimize waste and maximize resource utilization. These AI-driven solutions not only reduce food waste but also contribute to the overall sustainability of the food supply chain, ensuring that food resources are utilized efficiently and responsibly. In recent years, the concept of sustainable agriculture has gained significant traction as the world grapples with the challenges of feeding a growing population while minimizing environmental impact. Sustainable agriculture practices are those that aim to protect and enhance the health of the land, promote biodiversity, and ensure the economic viability of farming operations. In this subchapter, we will provide an overview of some key sustainable agriculture practices that are being increasingly adopted by farmers around the world.

One of the central tenets of sustainable agriculture is the use of integrated pest management (IPM) strategies to reduce reliance on chemical pesticides. IPM involves the careful monitoring of pest populations and the use of a combination of biological, cultural, and chemical control methods to manage pest outbreaks. By using IPM techniques, farmers can minimize the negative impacts of pesticides on the environment and human health while still effectively managing pest populations.

Another important aspect of sustainable agriculture is the promotion of soil health through practices such as crop rotation, cover cropping, and reduced tillage. These practices help to build soil organic matter, improve soil structure, and enhance nutrient cycling, resulting in healthier, more resilient soils that can support productive and sustainable agriculture in the long term.

Water conservation is also a key focus of sustainable agriculture practices, particularly in regions facing water scarcity. Farmers are increasingly adopting techniques such as drip irrigation, rainwater harvesting, and the use of drought-resistant crop varieties to minimize water use and maximize water efficiency in agricultural production. By conserving water resources, farmers can reduce their environmental footprint and ensure the long-term sustainability of their operations.

Sustainable agriculture practices are essential for ensuring the future of food production in a way that is environmentally responsible, economically viable, and socially equitable. By adopting practices such as integrated pest management, soil health promotion, and water conservation, farmers can contribute to a more sustainable food system that meets the needs of a growing population while protecting the health of the planet. It is imperative that nutritionists, farmers, food scientists, and food technologists work together to promote and support the adoption of sustainable agriculture practices to ensure a more resilient and sustainable food system for future generations.

32.1. Role of AI in Promoting Sustainability in Agriculture

In recent years, the role of artificial intelligence (AI) in promoting sustainability in agriculture has gained significant attention. This subchapter will delve into the various ways in which AI is revolutionizing the food industry and contributing to a more sustainable future. From precision agriculture to sustainable farming practices, AI is playing a crucial role in addressing the challenges facing the agriculture sector.

One of the key areas where AI is making a significant impact is in precision agriculture. By utilizing AI-powered technologies such as drones, sensors, and data analytics, farmers are able to monitor and manage their crops more effectively. This not only leads to increased productivity and reduced waste but also helps in the conservation of resources such as water and fertilizers. AI is enabling farmers to make data-driven decisions that optimize crop yields while minimizing environmental impact.

In addition to precision agriculture, AI is also driving automation in food processing and packaging. By implementing AI-driven technologies in food processing plants, manufacturers are able to streamline operations, reduce energy consumption, and minimize food waste. This not only benefits the environment but also improves the efficiency and safety of food production processes. AI is revolutionizing the way food is processed and packaged, making the industry more sustainable and environmentally friendly.

Furthermore, AI applications in sustainable farming practices are helping farmers adopt more eco-friendly techniques. From predictive analytics for weather forecasting to soil health monitoring, AI is providing farmers with valuable insights that enable them to make informed decisions that promote sustainability. By optimizing resource management and reducing the use of harmful chemicals, AI is helping farmers adopt practices that are not only portable but also environmentally responsible.

Moreover, AI-powered personalized nutrition and dietary recommendations are revolutionizing the way we approach food consumption. By analyzing individual health data and dietary preferences, AI algorithms are able to provide personalized recommendations that promote healthier eating habits. This not only benefits the individual but also contributes to a more sustainable food system by reducing food waste and promoting a more balanced diet . AI is enabling consumers to make more informed choices that are not only good for their health but also for the planet. the role of AI in promoting sustainability in agriculture is undeniable. From precision agriculture to sustainable farming practices, AI is transforming the food industry and leading the way towards a more sustainable future. By harnessing the power of AI- driven technologies, we can create a more efficient, eco-friendly, and ethical food system that benefits both people and the planet. As nutritionists, farmers, food scientists, and food technologists, it is crucial to embrace AI and its potential to revolutionize the way we produce, process, and consume food.

32.2. REMOTE SENSING AND IMAGING TECHNOLOGIES

Remote sensing and imaging technologies play a crucial role in the future of food production, especially in the realm of sustainable agriculture. These cutting-edge technologies allow farmers to gather valuable data about their crops and land from a distance, enabling them to make informed decisions that can increase crop yields, reduce environmental impact, and improve overall efficiency. By harnessing the power of remote sensing and imaging technologies, farmers can monitor crop health, detect diseases and pests early, optimize irrigation practices, and even predict yields with greater accuracy.

One of the key benefits of utilizing remote sensing and imaging technologies in agriculture is the ability to gather real-time data about crop conditions and environmental factors. This data can be used to create detailed maps of elds, identify areas of stress or nutrient deficiencies, and track changes in crop health over time. By analyzing this data, farmers can make more informed decisions about when and where to apply fertilizers, pesticides, and water, leading to more sustainable and environmentally friendly farming practices.

AI and its role in food production have revolutionized the way farmers utilize remote sensing and imaging technologies. Through the use of machine learning algorithms, AI applications can analyze vast amounts of data collected from sensors and cameras mounted on drones, satellites, and other platforms. This data can be used to create predictive models that help farmers optimize their operations, reduce waste, and increase productivity. By combining AI with remote sensing and imaging technologies, farmers can achieve higher levels of precision in their decision-making processes, leading to improved crop yields and reduced environmental impact.

In the realm of food processing and packaging, AI-driven automation is transforming the way food products are handled and packaged. By using computer vision systems, robotics, and machine learning algorithms, food processing plants can streamline their operations, increase efficiency, and reduce errors. AI-powered systems can identify defects in products, sort items based on quality criteria, and even optimize packaging processes to reduce waste. By leveraging AI-driven automation in food processing and packaging, companies can improve product quality, reduce costs, and enhance overall sustainability.

In conclusion, remote sensing and imaging technologies, when combined with AI, have the potential to revolutionize the way food is produced, processed, and packaged. By utilizing these advanced technologies, farmers can make more informed decisions about their crops, leading to higher yields and reduced environmental impact. In food processing and packaging, AI-driven automation can improve efficiency, reduce waste, and enhance product quality. As we look to the future of food production, it is clear that remote sensing and imaging technologies, along with AI, will play a vital role in shaping a more sustainable and efficient food system for generations to come.

32.3. SOIL MONITORING AND MANAGEMENT

Soil monitoring and management are crucial aspects of sustainable agriculture that can greatly benefit from the integration of AI technology . By utilizing AI-powered tools and algorithms, farmers, nutritionists, and food scientists can gather real-time data on soil health, nutrient levels, and moisture content, allowing for more accurate and efficient decision-making processes.

AI applications in precision agriculture have revolutionized the way farmers approach soil monitoring and management. Through the use of sensors, drones, and satellite imagery, farmers can collect data on soil composition, temperature, and moisture levels with unprecedented precision. This data can then be analyzed by AI algorithms to generate insights and recommendations for optimal soil management practices, such as fertilization schedules, irrigation plans, and crop rotation strategies.

AI-driven automation in food processing and packaging can also play a significant role in improving soil health and sustainability. By streamlining production processes and reducing waste, AI-powered systems can help minimize the environmental impact of food production while ensuring the quality and safety of the final product . Additionally, AI technologies can be used to optimize resource utilization, reduce energy consumption, and minimize carbon emissions, all of which can contribute to healthier soil and more sustainable farming practices.

AI applications in sustainable farming practices can further enhance soil monitoring and management efforts. By leveraging AI algorithms to analyze data from weather patterns, crop yields, and soil health indicators, farmers can make more informed decisions about land use, crop selection, and conservation practices. This holistic approach to farming can help preserve soil quality, protect natural ecosystems, and promote long-term sustainability in agriculture.

In conclusion, the integration of AI technology into soil monitoring and management practices holds great potential for improving soil health, sustainability, and productivity in agriculture. By harnessing the power of AI-driven tools and algorithms, nutritionists, farmers, and food scientists can make more informed decisions, optimize resource utilization, and reduce the environmental impact of food production. As we look towards the future of food production, AI will undoubtedly play a key role in shaping more sustainable and efficient farming practices.

32.4. CROP HEALTH MONITORING AND DISEASE DETECTION

Crop health monitoring and disease detection are critical aspects of sustainable agriculture that can greatly benefit from the use of AI technology . By leveraging AI tools, farmers, agronomists, and researchers can more effectively monitor the health of their crops, detect diseases early on, and implement timely interventions to prevent crop loss and ensure optimal yields.

One of the key advantages of using AI for crop health monitoring is its ability to analyze vast amounts of data from various sources, such as satellite imagery, weather data, and soil sensors, to provide real-time insights into the condition of crops. AI algorithms can detect subtle changes in plant health that may not be visible to the naked eye, allowing farmers to take proactive measures to address potential issues before they escalate.

In addition to monitoring crop health, AI can also play a crucial role in disease detection. By analyzing patterns in plant symptoms, AI algorithms can identify the presence of diseases or pests early on, enabling farmers to take targeted actions, such as applying specic treatments or implementing quarantine measures, to prevent the spread of the disease and minimize its impact on crop yields.

Furthermore, AI-driven technologies, such as machine learning and computer vision, can help automate the process of disease detection, making it faster, more accurate, and less labor-intensive for farmers. By deploying AI-powered drones or sensors in the eld, farmers can continuously monitor their crops for signs of disease, allowing for timely interventions and reducing the risk of crop loss.

Overall, the integration of AI in crop health monitoring and disease detection has the potential to revolutionize sustainable agriculture by enabling farmers to make more informed decisions, optimize resource allocation, and improve crop productivity. By embracing AI technologies, nutritionists, farmers, food scientists, and food technologists can contribute to the advancement of sustainable farming practices and ensure a more secure and resilient food supply for future generations.

32.5. ROBOTICS AND AUTOMATION IN FOOD PROCESSING

Robotics and automation have revolutionized the food processing industry, leading to increased efficiency, improved quality control, and reduced production costs. In the subchapter "Robotics and Automation in Food Processing" of our book "The Future of Food: AI and Its Role in Sustainable Agriculture," we delve into the innovative applications of artificial intelligence (AI) in streamlining food production processes. This chapter is essential reading for nutritionists, farmers, food scientists, and food technologists who are interested in harnessing the power of AI to enhance their operations.

One of the key benefits of incorporating robotics and automation in food processing is the ability to increase consistency and precision in manufacturing processes. By leveraging AI-driven technologies, food producers can optimize their production lines to ensure that each product meets stringent quality standards. This not only improves the overall quality of the end product but also reduces the risk of contamination and ensures consumer safety.

Furthermore, AI applications in precision agriculture have enabled farmers to monitor crop growth, soil health, and weather conditions with unprecedented accuracy. By using drones, sensors, and other AI-powered tools, farmers can make data-driven decisions that optimize crop yields and minimize resource wastage. This not only benefits the environment by reducing the use of pesticides and fertilizers but also improves the sustainability of farming practices .

In the realm of food processing and packaging, AI-driven automation has revolutionized the way products are manufactured, packaged, and distributed. Robotics and automated systems can perform repetitive tasks with speed and precision, reducing human error and increasing production output. This results in cost savings for food producers and ensures that products are delivered to consumers in a timely manner, enhancing customer satisfaction.

In conclusion, the integration of robotics and automation in food processing is reshaping the future of food production. By harnessing the power of AI, nutritionists, farmers, food scientists, and food technologists can optimize their operations, improve quality control, and reduce waste. As we continue to explore the potential of AI in sustainable agriculture, it is clear that robotics and automation will play a crucial role in shaping the future of the food industry.

32.6. AI APPLICATIONS IN QUALITY CONTROL AND ASSURANCE

In recent years, the integration of Artificial Intelligence (AI) technologies in various industries has revolutionized processes and operations, and the food production sector is no exception. AI applications in quality control and assurance play a crucial role in ensuring that the food we consume is safe, nutritious, and of high quality. This subchapter delves into the various ways in which AI is being utilized to enhance quality control and assurance in the food industry.

One of the key areas where AI is making a significant impact is in the detection of contaminants and pathogens in food products. By leveraging machine learning algorithms, food manufacturers can analyze vast amounts of data to identify potential risks and take corrective actions before they escalate into larger issues. This proactive approach not only helps in safeguarding consumer health but also minimizes the financial losses associated with product recalls and reputation damage.

AI-driven automation in food processing and packaging is another aspect that is transforming the quality control and assurance landscape. With the help of AI-powered systems, food manufacturers can streamline production processes, reduce human errors, and ensure consistency in product quality. By automating tasks such as sorting, labeling, and packaging, companies can improve efficiency and reduce the likelihood of defects or contamination in the final product.

Sustainable farming practices are increasingly becoming a focus for the food industry, and AI is playing a pivotal role in driving this shift towards more environmentally friendly and efficient methods. Through the use of AI-powered tools, farmers can optimize resource usage, minimize waste, and enhance crop yields. By analyzing data on weather patterns, soil conditions, and pest threats, AI systems can provide valuable insights that help farmers make informed decisions to improve productivity and sustainability.

In the realm of personalized nutrition and dietary recommendations, AI technologies are enabling food scientists and nutritionists to tailor dietary plans to individual needs and preferences. By analyzing data on a person's genetic makeup, lifestyle habits, and health goals, AI algorithms can generate personalized recommendations for optimal nutrition and wellness. This personalized approach not only enhances the effectiveness of dietary interventions but also empowers individuals to make informed choices about their food consumption. AI applications in quality control and assurance are revolutionizing the way food is produced, processed, and consumed. By leveraging AI technologies, the food industry can enhance safety, quality, and sustainability across the entire supply chain. As nutritionists, farmers, food scientists, and food technologists, it is imperative to embrace these advancements and collaborate to harness the full potential of AI in shaping the future of food production.

32.7. SMART PACKAGING SOLUTIONS ENABLED BY AI

In recent years, the integration of AI technology into various industries has revolutionized the way we approach problem-solving and decision- making processes. One area where AI has shown immense potential is in the realm of smart packaging solutions. By leveraging AI algorithms and machine learning capabilities, food manufacturers and packaging companies are able to create innovative packaging solutions that not only enhance the shelf life of food products but also improve overall sustainability and reduce food waste.

One of the key advantages of using AI in packaging solutions is the ability to monitor and track the condition of food products in real-time. Through the use of sensors and IoT devices, AI-powered packaging can detect changes in temperature, humidity, and other environmental factors that may impact the quality of the food. This allows for timely interventions to be made to prevent spoilage and ensure that the food remains fresh for a longer period of time. Additionally, AI can also optimize packaging designs to minimize the use of materials and reduce the carbon footprint of the packaging process.

Another important aspect of smart packaging solutions enabled by AI is the integration of smart labels and QR codes that provide consumers with valuable information about the food product. These labels can contain details such as the origin of the food, nutritional content, allergen information, and even cooking instructions. This level of transparency not only helps consumers make informed choices about their food purchases but also fosters trust and loyalty towards the brand. Furthermore, AI can analyze consumer preferences and feedback to tailor packaging designs and marketing strategies to better meet the demands of the target audience.

In the realm of sustainable agriculture, AI-powered packaging solutions play a crucial role in reducing food waste and promoting environmentally friendly practices. By utilizing predictive analytics and machine learning algorithms, packaging companies can optimize inventory management, transportation routes, and storage conditions to minimize losses and maximize efficiency. This not only benefits the bottom line of food manufacturers but also contributes to the overall sustainability of the food supply chain. Additionally, AI can help identify opportunities for recycling and composting packaging materials, further reducing the environmental impact of the packaging process.

Overall, the integration of AI technology into smart packaging solutions represents a significant step forward in the quest for more efficient, sustainable, and consumer- friendly food packaging. By leveraging the power of AI algorithms and machine learning capabilities, food manufacturers and packaging companies can create innovative packaging solutions that not only enhance the shelf life of food products but also improve overall sustainability and reduce food waste. As we continue to explore the potential of AI in the realm of food production and agriculture, smart packaging solutions enabled by AI will undoubtedly play a key role in shaping the future of the food industry.

32.8. PRECISION IRRIGATION AND WATER MANAGEMENT

Precision irrigation and water management play a crucial role in modern agriculture, especially in the face of increasing water scarcity and the need for sustainable farming practices. By utilizing AI technologies, farmers can optimize water usage and ensure that crops receive the right amount of water at the right time. This not only helps in conserving water resources but also improves crop yields and quality.

AI and its role in precision agriculture have revolutionized the way farmers approach irrigation and water management. Through the use of sensors, drones, and satellite imaging, farmers can monitor soil moisture levels, weather conditions, and crop health in real-time. This data is then analyzed by AI algorithms to provide insights and recommendations on when and where to irrigate, leading to more efficient water usage and better crop outcomes.

One of the key benefits of AI-driven automation in food processing and packaging is the ability to optimize water usage in these processes. By using AI-powered systems to monitor and control water usage, food manufacturers can reduce wastage and costs while maintaining high levels of food safety and quality. This not only benefits the bottom line but also contributes to sustainable water management practices in the food industry.

In sustainable farming practices, AI applications play a significant role in helping farmers make informed decisions about water usage and conservation. By analyzing data on soil health, weather patterns, and crop growth, AI can provide personalized recommendations on irrigation schedules, crop rotations, and soil management practices that optimize water usage and minimize environmental impact. This not only promotes sustainable farming practices but also ensures the long-term viability of agriculture.

Overall, the integration of AI technologies in precision irrigation and water management is a game-changer for the agriculture industry. By harnessing the power of AI to optimize water usage, farmers can improve crop yields, conserve water resources, and promote sustainable farming practices. This not only benefits the environment but also contributes to the production of healthier and more nutritious food for consumers.

32.9. CLIMATE PREDICTION AND ADAPTATION STRATEGIES

Climate prediction and adaptation strategies are crucial components in ensuring the sustainability of agriculture in the face of changing environmental conditions. With the advancements in AI technology, we now have powerful tools at our disposal to better understand climate patterns and develop effective adaptation strategies. This subchapter will discuss the importance of climate prediction and adaptation in the context of food production, and how AI can be leveraged to enhance these efforts.

One of the key challenges facing farmers and food producers today is the unpredictability of weather patterns due to climate change. By utilizing AI algorithms that can analyze vast amounts of data, we can improve our ability to predict future climate trends with greater accuracy. This information can then be used to develop adaptive strategies that help mitigate the impact of extreme weather events on crops and livestock.

In the realm of precision agriculture, AI plays a crucial role in optimizing resource use and maximizing yields in the face of changing climate conditions. By integrating AI-driven technologies such as drones, sensors, and data analytics, farmers can monitor crop health and soil conditions in real-time, allowing for timely interventions to address potential issues before they escalate. This proactive approach not only improves productivity but also reduces the environmental impact of agriculture.

In food processing and packaging, AI-driven automation systems can help food manufacturers adapt to changing climate conditions by optimizing production processes and reducing energy consumption. By utilizing AI algorithms to streamline operations and minimize waste, companies can improve efficiency and reduce their carbon footprint. Additionally, AI can help identify areas where packaging materials can be optimized for greater sustainability, further reducing the environmental impact of food production.

Climate prediction and adaptation strategies are essential for ensuring the resilience of food production systems in the face of climate change. By harnessing the power of AI in areas such as precision agriculture, food processing, and packaging, we can develop more sustainable practices that not only increase productivity but also reduce environmental impact. As nutritionists, farmers, food scientists, and food technologists, it is imperative that we embrace AI-driven solutions to help create a more sustainable future for food production.

32.10. SUSTAINABLE PEST AND WEED MANAGEMENT WITH AI

In recent years, the use of AI in agriculture has revolutionized the way farmers manage pests and weeds on their farms. This technology has enabled farmers to implement more sustainable and environmentally friendly practices that reduce the need for harmful chemicals and pesticides. By harnessing the power of AI, farmers can now accurately identify and target specific pests and weeds, allowing for more precise and efficient control measures.

One of the key advantages of using AI for pest and weed management is its ability to analyze vast amounts of data in real-time. This means that farmers can quickly identify emerging pest and weed threats and take proactive measures to address them before they become a problem. By utilizing AI-powered sensors and drones, farmers can monitor their fields more effectively and make data-driven decisions on when and where to apply pest and weed control measures.

Furthermore, AI technology can help farmers optimize their use of resources by only applying pest and weed control measures when and where they are needed. This not only reduces the environmental impact of agriculture but also saves farmers time and money in the long run. By utilizing AI-driven algorithms, farmers can create personalized pest and weed management plans that take into account factors such as crop type, soil health, and weather conditions.

Overall, the integration of AI into pest and weed management practices is a game- changer for sustainable agriculture. By leveraging the power of AI, farmers can reduce their reliance on harmful chemicals, minimize the environmental impact of agriculture, and improve the overall health and productivity of their crops. As we look towards the future of food production, AI will continue to play a crucial role in helping farmers adopt more sustainable practices that benefit both the environment and consumer health.

32.11. DATA ANALYSIS AND PERSONALIZED NUTRITION PLANS

In the realm of nutrition and personalized nutrition plans, data analysis plays a crucial role in providing tailored dietary recommendations to individuals. By utilizing AI technologies, nutritionists can analyze vast amounts of data to understand individual health needs, dietary preferences, and lifestyle factors. This allows for the creation of personalized nutrition plans that are tailored to each individual's unique requirements, leading to better health outcomes and overall well-being.

For farmers and food scientists, data analysis and personalized nutrition plans can also be valuable tools in optimizing food production and ensuring the delivery of nutritious and sustainable food options. By leveraging AI technologies, farmers can analyze data on soil health, weather patterns, and crop yields to make informed decisions on planting, harvesting, and distribution. This can help improve crop yields, minimize waste, and ensure the availability of nutritious food options for consumers. In the field of food technology, data analysis is also revolutionizing the way food is processed and packaged.

AI-driven automation systems can analyze data on food quality, safety, and nutritional content to optimize production processes and ensure the delivery of high-quality, nutritious food products. This not only improves efficiency and reduces costs but also ensures that consumers have access to safe and nutritious food options.

Furthermore, AI applications in sustainable farming practices are helping farmers adopt environmentally friendly practices that minimize the impact on the planet. By analyzing data on soil health, water usage, and crop rotation, AI technologies can help farmers make informed decisions on sustainable farming practices that help conserve natural resources, reduce greenhouse gas emissions, and promote biodiversity.

Overall, data analysis and personalized nutrition plans are transforming the future of food by providing tailored dietary recommendations, optimizing food production processes, and promoting sustainable farming practices. By harnessing the power of AI technologies, nutritionists, farmers, food scientists, and food technologists can work together to create a more sustainable and nutritious food system for future generations.

32.12. AI-POWERED MEAL PLANNING AND RECIPE SUGGESTIONS

In recent years, Artificial Intelligence (AI) has revolutionized various industries, including the food and agriculture sector. One of the most exciting applications of AI in this realm is AI-powered meal planning and recipe suggestions. This technology leverages machine learning algorithms to analyze vast amounts of data on nutritional value, dietary preferences, and ingredient availability to offer personalized meal plans and recipe recommendations to individuals.

For nutritionists, AI-powered meal planning and recipe suggestions provide a valuable tool for creating customized dietary plans for their clients. By inputting specific dietary requirements and preferences into the AI system, nutritionists can generate meal plans that meet their clients' nutritional needs while also accommodating their tastes and lifestyle. This not only streamlines the meal planning process but also ensures that clients are receiving a well-balanced diet tailored to their individual needs.

For farmers and food scientists, AI-powered meal planning and recipe suggestions offer insights into consumer preferences and trends, allowing them to adjust their production and supply chains accordingly. By analyzing data on popular recipes and ingredients, farmers can make informed decisions about what crops to plant and how to optimize their yields. Similarly, food scientists can use AI-generated recipe suggestions to develop innovative products that cater to changing consumer tastes and demands. Food technologists can also benefit from AI-powered meal planning and recipe suggestions by using these tools to optimize food processing and packaging techniques. By incorporating AI algorithms into their operations, food technologists can improve the efficiency and quality of their products, ultimately reducing waste and enhancing sustainability throughout the supply chain. This not only benefits the environment but also helps companies save on costs and improve their bottom line.

Overall, AI-powered meal planning and recipe suggestions have the potential to revolutionize the way we approach food production and consumption. By harnessing the power of AI to personalize nutrition, reduce waste, and optimize food production processes, we can create a more sustainable and efficient food system that benefits both individuals and the planet. As we continue to explore the possibilities of AI in the food and agriculture sector, it is essential for nutritionists, farmers, food scientists, and food technologists to embrace these technologies and leverage them for the betterment of our food system.

32.13. NUTRITIONAL MONITORING AND FEEDBACK SYSTEMS

In the realm of sustainable agriculture, the integration of AI has revolutionized the way we monitor and provide feedback on the nutritional content of our food. Nutritional monitoring and feedback systems play a crucial role in ensuring that both farmers and consumers have access to accurate and up-to-date information about the nutritional value of the food they produce and consume. By leveraging AI technologies, we can now track and analyze the nutritional content of crops, livestock, and processed foods in real-time, enabling us to make more informed decisions about our diets and farming practices.

One of the key applications of AI in nutritional monitoring and feedback systems is in precision agriculture. Through the use of sensors, drones, and other advanced technologies, farmers can collect data on soil quality, crop health, and weather conditions to optimize their farming practices and maximize crop yields. AI algorithms can then analyze this data to provide recommendations on the use of fertilizers, pesticides, and other inputs to ensure that crops are grown in a sustainable and nutritionally rich manner.

In addition to precision agriculture, AI-driven automation is also being used in food processing and packaging to ensure that the nutritional content of foods is preserved throughout the production process. By using AI-powered systems to monitor and control the temperature, humidity, and other factors that can affect the nutritional value of food, manufacturers can ensure that consumers receive products that are both safe and nutritious. This level of automation not only improves the quality of our food but also reduces waste and improves the overall sustainability of our food supply chain.

Furthermore, AI applications in sustainable farming practices are helping to address some of the biggest challenges facing the agricultural industry today, such as water scarcity, soil degradation, and climate change. By using AI to optimize irrigation systems, monitor soil health, and predict weather patterns, farmers can reduce their environmental impact while also improving the nutritional content of their crops. This not only benefits the environment but also ensures that consumers have access to healthier and more sustainable food options.

Overall, the integration of AI into nutritional monitoring and feedback systems has the potential to revolutionize the way we produce, process, and consume food. By leveraging the power of AI to track and analyze the nutritional content of our food, we can make more informed decisions about our diets, improve the sustainability of our food supply chain, and ultimately create a healthier and more environmentally friendly food system for future generations.

32.14. AI-POWERED FOOD WASTE REDUCTION AND MANAGEMENT SOLUTIONS

Smart inventory management systems are revolutionizing the way food is produced, processed, and distributed in the agricultural industry. By leveraging AI technology, farmers, food scientists, and food technologists are able to optimize their supply chain operations, reduce waste, and improve overall efficiency. These systems use advanced algorithms to track inventory levels, predict demand, and automate replenishment processes, ensuring that the right amount of food is available at the right time.

One of the key benefits of smart inventory management systems is their ability to prevent overstocking and understocking of food products. By analyzing historical sales data, weather patterns, and other relevant factors, these systems can accurately forecast demand and adjust inventory levels accordingly. This not only helps to reduce food waste and minimize losses but also ensures that customers have access to fresh and high- quality products at all times.

AI-powered inventory management systems also play a crucial role in sustainable farming practices by enabling farmers to monitor and optimize their resource use. By tracking the use of water, fertilizers, and pesticides, farmers can reduce their environmental impact and improve the overall sustainability of their operations. Additionally, these systems can help farmers identify areas of improvement and implement more efficient farming practices, ultimately leading to higher yields and reduced costs.

In the food processing and packaging industry, AI-driven automation is revolutionizing the way products are manufactured and distributed. By incorporating AI technology into production lines, companies can improve efficiency, reduce errors, and enhance product quality. These systems can also help to streamline inventory management processes, ensure timely delivery of products, and minimize waste throughout the supply chain. Overall, smart inventory management systems are playing a crucial role in shaping the future of food production and distribution. By leveraging AI technology, nutritionists, farmers, food scientists, and food technologists can optimize their operations, reduce waste, and improve sustainability. As the agricultural industry continues to evolve, these systems will be instrumental in driving innovation and ensuring a more efficient and sustainable food supply chain.

32.15. PREDICTIVE ANALYTICS FOR WASTE REDUCTION

Predictive analytics for waste reduction is a crucial aspect of sustainable agriculture and food production. By harnessing the power of AI, nutritionists, farmers, food scientists, and food technologists can optimize their processes to minimize food waste and maximize efficiency. This subchapter delves into the various applications of predictive analytics in reducing waste throughout the food supply chain.

AI plays a significant role in predicting and preventing food waste at every stage of production. By analyzing data from sensors, cameras, and other sources, AI algorithms can identify patterns and trends that indicate potential sources of waste. This allows food producers to take proactive measures to address these issues before they escalate, saving both resources and valuable food items.

One of the key benefits of using predictive analytics for waste reduction is the ability to optimize inventory management. By accurately forecasting demand and supply, food producers can minimize overstocking and understocking, which are common causes of food waste. AI algorithms can also analyze factors such as shelf life, storage conditions, and transportation logistics to ensure that food products reach consumers in optimal condition.

Furthermore, predictive analytics can help food producers identify opportunities for process optimization and efficiency improvements. By analyzing data on production cycles, equipment performance, and resource utilization, AI algorithms can pinpoint areas where waste can be reduced through better planning and decision-making. This not only leads to cost savings but also contributes to a more sustainable and environmentally friendly food production system. predictive analytics for waste reduction is a powerful tool that can help nutritionists, farmers, food scientists, and food technologists achieve their goals of sustainability and efficiency in food production. By leveraging AI technology to analyze data and make informed decisions, food producers can minimize waste, improve resource utilization, and ultimately contribute to a more sustainable future for the food industry. It is essential for professionals in the field to embrace these technologies and incorporate them into their practices to ensure a more efficient and responsible food supply chain.

32.16. AI-ENABLED FOOD RESCUE AND REDISTRIBUTION PLATFORMS

AI-enabled food rescue and redistribution platforms are revolutionizing the way we approach food waste management and sustainable food distribution. These innovative platforms utilize artificial intelligence to efficiently match surplus food from restaurants, grocery stores, and farms with organizations and individuals in need. By leveraging AI algorithms, these platforms are able to quickly identify surplus food items, determine their quality and expiration dates, and facilitate their redistribution to ensure they are consumed before they go to waste. Nutritionists, farmers, food scientists, and food technologists can all benefit from AI- enabled food rescue and redistribution platforms . Nutritionists can access fresh, nutritious food for their clients in need, while farmers can reduce food waste on their farms by donating surplus produce. Food scientists can use these platforms to conduct research on food waste reduction and sustainable food distribution practices, while food technologists can leverage AI to improve the efficiency of food processing and packaging.

One of the key advantages of AI-enabled food rescue and redistribution platforms is their ability to scale and adapt to changing needs and conditions. By continuously analyzing data on food surplus and demand, these platforms can adjust their algorithms to match supply with demand in real-time. This flexibility allows for more efficient and effective redistribution of surplus food, ultimately reducing food waste and ensuring that more people have access to fresh, nutritious food.

In addition to reducing food waste and improving food distribution, AI-enabled food rescue and redistribution platforms also have the potential to reduce carbon emissions and environmental impact associated with food production and transportation. By diverting surplus food from landfills and incinerators, these platforms help to reduce methane emissions and minimize the environmental footprint of the food industry. Furthermore, by facilitating the redistribution of surplus food to those in need, these platforms contribute to social and economic sustainability by addressing food insecurity and promoting food access for all. Overall, AI-enabled food rescue and redistribution platforms are a powerful tool in the fight against food waste and in promoting sustainable food distribution practices. By harnessing the power of artificial intelligence, these platforms offer a scalable, efficient, and effective solution to the challenges of food waste management and food distribution. As nutritionists, farmers, food scientists, and food technologists continue to explore the potential of AI in food production, distribution, and waste reduction, these platforms will play an increasingly important role in shaping the future of food sustainability.

32.16.1. Future Prospects and Challenges in AI and Sustainable Agriculture

In the rapidly evolving landscape of agriculture and food production, opportunities for innovation and growth abound as AI continues to play a pivotal role in shaping the future of food. For nutritionists, farmers, food scientists, and food technologists, the integration of AI technologies offers a myriad of possibilities to enhance efficiency, sustainability, and productivity across the entire food supply chain.

One of the key areas where AI is driving innovation and growth is in food production. By leveraging AI-powered analytics and predictive modeling, farmers can optimize crop yields, reduce waste, and mitigate environmental impact. AI applications in precision agriculture, such as drone technology and sensor-based monitoring systems, enable farmers to make data-driven decisions that lead to more efficient resource allocation and improved crop management practices.

In food processing and packaging, AI-driven automation is revolutionizing the industry by streamlining operations, enhancing quality control, and increasing production throughput. By implementing AI-powered algorithms for quality assurance and predictive maintenance, food manufacturers can ensure product consistency, reduce downtime, and improve overall efficiency. Additionally, AI technologies enable real-time monitoring of food safety standards and compliance with regulatory requirements, ensuring the highest level of consumer protection.

Sustainable farming practices are also benefiting from AI applications, as advanced algorithms and machine learning techniques enable farmers to optimize resource usage, minimize environmental impact, and promote biodiversity. By integrating AI-driven solutions for soil health management, pest control, and water conservation, farmers can cultivate healthier crops while preserving natural ecosystems. AI-powered tools also facilitate the adoption of regenerative agriculture practices, leading to improved soil fertility, carbon sequestration, and long-term sustainability.

In the realm of personalized nutrition and dietary recommendations, AI is revolutionizing the way individuals access and interpret nutritional information.

By leveraging AI algorithms for dietary analysis, meal planning, and food tracking, nutritionists can provide tailored recommendations that cater to each individual's unique needs and preferences. AI-powered platforms also enable consumers to make informed choices about their diet, leading to healthier lifestyles and improved well-being.

Furthermore, AI technologies offer innovative solutions for food waste reduction and management, addressing one of the most pressing challenges facing the food industry today. By implementing AI-driven analytics for inventory management, demand forecasting, and supply chain optimization, food businesses can minimize waste, reduce costs, and enhance sustainability. AI-powered solutions also enable real-time monitoring of food spoilage, expiration dates, and storage conditions, ensuring that food reaches consumers in optimal quality while minimizing environmental impact.

32.17. ETHICAL AND REGULATORY CONSIDERATIONS IN AI APPLICATIONS

In the realm of AI applications in food production, there are several ethical and regulatory considerations that must be taken into account. As nutritionists, farmers, food scientists, and food technologists delve deeper into the possibilities of AI in agriculture, it is crucial to uphold ethical standards and comply with regulatory guidelines to ensure the responsible use of this technology.

One of the key ethical considerations in AI applications in food production is the potential impact on human health and safety. As AI-driven automation becomes more prevalent in precision agriculture and food processing, it is important to ensure that these technologies do not compromise the quality or safety of the food being produced. This includes monitoring the accuracy of AI algorithms, ensuring proper maintenance of AI systems, and conducting thorough risk assessments before implementing AI solutions in food production processes.

Another ethical consideration in AI applications in food production is the potential for job displacement. As AI-driven automation becomes more widespread in food processing and packaging, there is a risk that human workers may be replaced by machines. It is important for stakeholders in the food industry to consider the social implications of AI adoption and to prioritize the well-being of workers who may be affected by these changes.

From a regulatory standpoint, there are also considerations that must be taken into account when implementing AI applications in food production. Regulatory bodies may have specific guidelines and requirements for the use of AI in agriculture, and it is essential for nutritionists, farmers, food scientists, and food technologists to stay informed and compliant with these regulations. This includes obtaining necessary permits, certifications, and approvals before deploying AI solutions in food production processes.

Ethical and regulatory considerations play a crucial role in the responsible implementation of AI applications in food production. By upholding ethical standards, prioritizing human health and safety, and complying with regulatory guidelines, nutritionists, farmers, food scientists, and food technologists can harness the power of AI to drive sustainable agriculture and food production practices. It is essential for stakeholders in the food industry to work together to ensure that AI technologies are used in a way that benefits both the industry and society as a whole.

The future of AI in sustainable agriculture holds great promise for addressing the challenges facing the global food system. As we have seen throughout this book, AI has the potential to revolutionize food production, from precision agriculture to food processing and packaging. By harnessing the power of AI technologies, nutritionists, farmers, food scientists, and food technologists can work together to create a more efficient and sustainable food supply chain.

One key area where AI can make a significant impact is in sustainable farming practices. AI-powered solutions can help farmers optimize their use of resources, reduce waste, and minimize environmental impact. By using data-driven insights and predictive analytics, farmers can make more informed decisions about crop management, water usage, and pest control, leading to higher yields and healthier ecosystems.

Furthermore, AI-driven automation in food processing and packaging can streamline operations, improve efficiency, and reduce costs. With the help of AI technologies such as robotics and machine learning, food manufacturers can increase productivity, ensure food safety, and meet the growing demand for nutritious and sustainable food products.

Another important application of AI in sustainable agriculture is personalized nutrition and dietary recommendations. By analyzing data from individual consumers, AI algorithms can provide personalized dietary advice, tailored to their specific nutritional needs and health goals. This can help people make better food choices, improve their overall health, and reduce the risk of chronic diseases.

CONCLUSION

Business as usual won't protect the future of our food supply—or the well-being of the farmers and communities that produce it. We need to take concrete steps to prepare for climate impacts on agriculture and to reduce both their severity and our vulnerability to them.

We also need to remember that climate change risks aren't distributed equally—and neither are the pathways to climate adaptation. Public policies and institutional practices have long denied communities of color, low-income groups, and tribal communities access to critical resources and decision-making processes, leaving them with fewer options and more risk in the face of climate impacts. So it's crucial to ensure that these communities have a voice in shaping our adaptation strategies.

Our farms and farm communities don't have to be sitting ducks for climate impacts. Forward-looking farmers and scientists are finding new, climate-resilient ways to produce our food:

Build healthier, "spongier" soils through practices—such as planting cover crops and deep-rooted perennials—that increase soil's capacity to soak up heavy rainfall and hold water for dry periods;

Make farms stronger by redesigning them as diverse agroecosystems—incorporating trees and native perennials, reducing dependence on fertilizers and pesticides, and reintegrating crops and livestock;

Develop new crop varieties, livestock breeds, and farm practices specifically designed to help farmers adapt to evolving climate realities.

Finally, whatever we do to help farmers adapt to climate change, we still face the urgent need and obligation to reduce the source of the problem as far and as fast as we can. This means bringing net emissions of heat-trapping gases down to zero, and doing it soon.

Fortunately, our farm and food system can be an important part of the solution, both by reducing emissions at every stage of the food production and distribution process, and by building agroecosystems that can sequester (store) more carbon.

BIBLIOGRAPHY

1. Abdulwaliyu, I., Okoduwa, S. I. R., Arekemase, S. O., et al. (2023). Impact of security challenges on food and nutrition security in Nigeria: The role of food production focus. *Malaysian Journal of Sustainable Agriculture, 7*(1), 45–51.
2. Abou-El-Hassan, S., Abd Elwanis, M., & El-Shinawy, M. Z. (2017). Application of compost and vermicompost as substitutes for mineral fertilizers to produce green beans. *Egyptian Journal of Horticulture, 44*(2), 155–163.
3. Adela, F. A., Aurbacher, J., & Abebe, G. K. (2019). Small-scale irrigation scheme governance–poverty nexus: Evidence from Ethiopia. *Food Security, 11*(4), 897–913.
4. Adeleke, B. S., Babalola, O. O., & Glick, B. R. (2021). Plant growth-promoting root-colonizing bacterial endophytes. *Rhizosphere, 20*, 100433.
5. Adenle, A. A., Wedig, K., & Azadi, H. (2019). Sustainable agriculture and food security in Africa: The role of innovative technologies and international organizations. *Technology in Society, 58*, 101143.
6. Adimassu, Z., Langan, S., & Johnston, R. (2016). Understanding determinants of farmers' investments in sustainable land management practices in Ethiopia: Review and synthesis. *Environment, Development and Sustainability, 18*(4), 1005–1023.
7. Adomako, M. O., Gao, F. L., Li, J. M., Du, D. L., Xue, W., & Yu, F. H. (2020). Effects of soil nutrient heterogeneity and parasitic plant infection on an experimental grassland community. *Flora, 271*, 151666.

8. Adomako, M. O., Xue, W., Du, D. L., & Yu, F. H. (2021). Soil biota and soil substrates influence responses of the rhizomatous clonal grass *Leymus chinensis* to nutrient heterogeneity. *Plant and Soil, 465*, 19–29.

9. Adugna, E., Ermias, A., Mekonnen, A., & Mihret, D. (2014). The role of small scale irrigation in poverty reduction. *Journal of Development and Agricultural Economics, 6*(1), 12–21.

10. Agegnehu, G., Nelson, P. N., & Bird, M. I. (2016). The effects of biochar, compost and their mixture and nitrogen fertilizer on yield and nitrogen use efficiency of barley grown on a Nitisol in the highlands of Ethiopia. *Science of the Total Environment, 569–570*, 869–879.

11. Aleksandrowicz, L., Green, R., Joy, E. J., Smith, P., & Haines, A. (2016). The impacts of dietary change on greenhouse gas emissions, land use, water use, and health: A systematic review. *PLoS ONE, 11*, e0165797.

12. Alexandre, P., Brown, C., Arneth, A., Dias, C., Finnigan, J., Moran, D., et al. (2017). Could consumption of insects, cultured meat or imitation meat reduce global agricultural land use? *Global Food Security, 15*, 22–32.

13. Alori, E. T., Glick, B. R., & Babalola, O. O. (2017). Microbial phosphorus solubilization and its potential for use in sustainable agriculture. *Frontiers in Microbiology, 8*, 971.

14. Altieri, M. A. (2018). *Agroecology: The Science of Sustainable Agriculture*. CRC Press.

15. Álvarez, C., Navarro, J. A., Molina-Heredia, F. P., & Mariscal, V. (2020). Endophytic colonization of rice (*Oryza sativa* L.) by the symbiotic strain *Nostoc punctiforme* PCC 73102. *Molecular Plant-Microbe Interactions, 33*(8), 1040–1045.

16. Amede, T. (2015). Technical and institutional attributes constraining the performance of small-scale irrigation in Ethiopia. *Water Resources and Rural Development, 6*, 78–91.

17. Andresen, E., Peiter, E., & Kupper, H. (2018). Trace metal metabolism in plants. *Journal of Experimental Botany, 69*(5), 909–954.

18. Andrews, M., & Andrews, M. E. (2017). Specificity in legume-rhizobia symbioses. *International Journal of Molecular Sciences, 18*(4), 705.

19. Andrews, S., et al. (2016). *Soil Health, Crop Health, and Food Quality*. University of California Agriculture and Natural Resources.

20. Appavu, B., Vanatta, L., Condie, J., Kerrigan, J. F., & Jarrar, R. (2016). Ketogenic diet treatment for pediatric super-refractory status epilepticus. *Seizure, 41*, 62–65.

21. Aroney, S. T. N., Poole, P. S., & Sánchez-Cañizares, C. (2021). Rhizobial chemotaxis and motility systems at work in the soil. *Frontiers in Plant Science, 12*, 725338.

22. Asfaw, S., Shiferaw, B., Simtowe, F., & Lipper, L. (2012). Impact of modern agricultural technologies on smallholder welfare: Evidence from Tanzania and Ethiopia. *Food Policy, 37*(3), 283–295.
23. Asner, G. P., Elmore, A. J., Olander, L. P., Martin, R. E., & Harris, T. (2004). Grazing systems, ecosystem responses, and global change. *Annual Review of Environment and Resources, 29*, 261–299.
24. Aulakh, M. S., & Malhi, S. S. (2005). Interactions of nitrogen with other nutrients and water: Effect on crop yield and quality, nutrient use efficiency, carbon sequestration, and environmental pollution. *Advances in Agronomy, 86*, 341–409.
25. Awaad, H., Abu-Hashim, M., & Negm, A. (2021). *Mitigating Environmental Stresses for Agricultural Sustainability in Egypt* (Vol. 16). Springer.
26. Aziz, M., Nadipalli, R. K., Xie, X., Sun, Y., Surowiec, K., Zhang, J. L., & Paré, P. W. (2016). Augmenting sulfur metabolism and herbivore defense in *Arabidopsis* by bacterial volatile signaling. *Frontiers in Plant Science, 7*, 458.
27. Azzam, A. (2020). Is the world converging to a 'Western diet'? *Public Health Nutrition, 24*, 309–317.
28. Bai, Z. G., Dent, D. L., Olsson, L., & Schaepman, M. E. (2008). Proxy global assessment of land degradation. *Soil Use and Management, 24*, 223–234.
29. Barberon, M., Dubeaux, G., Kolb, C., Isono, E., Zelazny, E., & Vert, G. (2014). Polarization of iron-regulated transporter 1 (IRT1) to the plant-soil interface plays crucial role in metal homeostasis. *PNAS USA, 111*(22), 8293–8298.
30. Baslam, M., Garmendia, I., & Goicoechea, N. (2013). The arbuscular mycorrhizal symbiosis can overcome reductions in yield and nutritional quality in greenhouse-lettuces cultivated at inappropriate growing seasons. *Scientia Horticulturae*, *164*, 145–154. https://doi.org/10.1016/j.scienta.2013.09.021.
31. Beltran-Peña, A., Rosa, L., & D'Odorico, P. (2020). Global food self-sufficiency in the 21st century under sustainable intensification of agriculture. *Environmental Research Letters*, *15*, ab9388.
32. Benson, D. R., & Silvester, W. B. (1993). Biology of Frankia strains, actinomycete symbionts of actinorhizal plants. *Microbiological Reviews*, *57*, 293–319.
33. Berhane, A. (2018). Climate change and variability impacts on agricultural productivity and food security. *Climatology Weather Forecast*, *6*(240), 25.

34. Bertol, I., Luciano, R. V., Bertol, C., & Bagio, B. (2017). Nutrient and organic carbon losses, enrichment rate, and cost of water erosion. *Ecosystem and Sustainable Agriculture*, *2*, 1–15.
35. Birch, C. P. D., & Hutchings, M. J. (1994). Exploitation of patchily distributed soil resources by the clonal herb Glechoma hederacea. *Journal of Ecology*, *82*, 653–664.
36. Birol, E., Meenakshi, J. V., Oparinde, A., Perez, S., & Tomlins, K. (2015). Developing country consumers' acceptance of biofortified foods: A synthesis. *Food Security*, *7*(3), 555–568.
37. Blair, N., Faulkner, R. D., Till, A. R., & Poulton, P. R. (2006). Long-term management impacts on soil C, N and physical fertility: Part I: Broadbalk experiment. *Soil and Tillage Research*, *91*, 30–38.
38. Bliss, K. M., Jones, R. H., Mitchell, R. J., & Mou, P. P. (2002). Are competitive interactions influenced by spatial nutrient heterogeneity and root foraging behavior? *New Phytologist*, *154*, 409–417.
39. Bouis, H. E., Hotz, C., McClafferty, B., Meenakshi, J. V., & Pfeiffer, W. H. (2011). Biofortification: A new tool to reduce micronutrient malnutrition. *Food and Nutrition Bulletin*, *32*(1 Suppl), S31–S40.
40. Brady, N. C., & Weil, R. R. (2008). *The Nature and Properties of Soils*. Prentice Hall.
41. Braga, L. P. P., Spor, A., Kot, W., Breuil, M. C., Hansen, L. H., Setubal, J. C., & Philippot, L. (2020). Impact of phages on soil bacterial communities and nitrogen availability under different assembly scenarios. *Microbiome*, *8*, 52.
42. Brevik, E. C., & Sauer, T. J. (2015). The past, present, and future of soils and human health studies. *SOIL*, *1*, 35–46.
43. Brinton, W. F. (2015). *Compost Utilization in Horticultural Cropping Systems*. CRC Press.
44. Brock, C., Oltmanns, M., Matthes, C., Schmehe, B., Schaaf, H., Burghardt, D., et al. (2021). Compost as an option for sustainable crop production at low stocking rates in organic farming. *Agronomy*, *11*(6), 1078.
45. Cahill, J. F., McNickle, G. G., Haag, J. J., Lamb, E. G., Nyanumba, S. M., & St. Clair, C. C. (2010). Plants integrate information about nutrients and neighbors. *Science*, *328*, 1657.
46. Cai, A., Chang, N., Zhang, W., Liang, G., Zhang, X., Hou, E., Jiang, L., Chen, X., Xu, M., & Luo, Y. (2019). The spatial patterns of litter turnover time in Chinese terrestrial ecosystems. *European Journal of Soil Science*, *71*, 856–867.
47. Carpenter, C. L. (2013). Food addiction and obesity. In C. Berdanier, D.

Dwyer, & D. Heber (Eds.), *CRC Handbook of Nutrition and Food* (3rd ed.). CRC Press.

48. Carrillo-Castañeda, G., Muños, J. J., Peralta-Videa, J., Gomez, E., Tiemannb, K., Duarte-Gardea, M., et al. (2002). Alfalfa growth promotion by bacteria grown under iron limiting conditions. *Advances in Environmental Research*, *6*(3), 391–399. https://doi.org/10.1016/S1093-0191(02)00054-0.
49. Carvalho, F. P. (2006). Agriculture, pesticides, food security and food safety. *Environmental Science and Policy*, *9*, 685–692.
50. Cassidy, E. S., West, P. C., Gerber, J. S., & Foley, J. A. (2013). Redefining agricultural yields: From tonnes to people nourished per hectare. *Environmental Research Letters*, *8*, 34015.
51. Castulo-Rubio, D. Y., Alejandre-Ramírez, N. A., Del Carmen, O. M. M., Santoyo, G., Macías-Rodríguez, L. I., & Valencia-Cantero, E. (2015). Volatile organic compounds produced by the rhizobacterium Arthrobacter agilis UMCV2 modulate *Sorghum bicolor* (strategy II plant) morphogenesis and *SbFRO1* transcription *in vitro*. *Journal of Plant Growth Regulation*, *34*(3), 611–623. https://doi.org/10.1007/s00344-015-9495-8.
52. Centers for Disease Control and Prevention. (2022). Adult obesity facts. Retrieved from: https://www.cdc.gov/obesity/data/adult.html (accessed on 17 October 2024).
53. Chai, B. C., van der Voort, J. R., Grofelnik, K., Eliasdottir, H. G., Klöss, I., & Perez-Cueto, F. J. A. (2019). Which diet has the least environmental impact on our planet? A systematic review of vegan, vegetarian, and omnivorous diets. *Sustainability*, *11*, 4110. https://doi.org/10.3390/su11154110.
54. Chandra, P. (2019). Soil–microbes–plants: Interactions and ecological diversity. In A. Varma, S. Tripathi, & R. Prasad (Eds.), *Plant Microbe Interface* (pp. 67–81). Springer.
55. Chauhan, S., Thakur, R., & Sharma, G. (2008). Nickel: Its availability and reactions in soil. *Journal of Industrial Pollution Control*, *24*(1), 1–8.
56. Checcucci, A., & Marchetti, M. (2020). The rhizosphere talk show: The rhizobia on stage. *Frontiers in Agronomy*, *2*, 591494.
57. Chen, D., Ali, A., Yong, X. H., Lin, C. G., Niu, X. H., Cai, A. M., Dong, B. C., Zhou, Z. X., Wang, Y. J., & Yu, F. H. (2019). A multi-species comparison of selective placement patterns of ramets in invasive alien and native clonal plants to light, soil nutrient and water heterogeneity. *Science of the Total Environment*, *657*, 1568–1577.
58. Chiba, M., Nakane, K., & Komatsu, M. (2019). Westernized diet is the

most ubiquitous environmental factor in inflammatory bowel disease. *The Permanente Journal*, *23*, 18–107. https://doi.org/10.7812/TPP/18-107.

59. Cleveland, C. C., Townsend, A. R., Schimel, D. S., Fisher, H., Howarth, R. W., Hedin, L. O., Perakis, S. S., Latty, E. F., Von Fischer, J. C., Elseroad, A., et al. (1999). Global patterns of terrestrial biological nitrogen (N2) fixation in natural ecosystems. *Glob. Biogeochem. Cycles*, *13*, 623–645.
60. Coccina, A., Cavagnaro, T. R., Pellegrino, E., Ercoli, L., Mclaughlin, M. J., & Watts-Williams, S. J. (2019). The mycorrhizal pathway of zinc uptake contributes to zinc accumulation in barley and wheat grain. *BMC Plant Biol.*, *19*(1), 1–14.
61. Coleine, C., Stajich, J. E., & Selbmann, L. (2022). Fungi are key players in extreme ecosystems. *Trends Ecol. Evol.*, *37*, 517–528.
62. Condro, A. A., Setiawan, Y., Prasetyo, L. B., Pramulya, R., & Siahaan, L. (2020). Retrieving the national main commodity maps in Indonesia based on high-resolution remotely sensed data using cloud computing platform. *LAND*, *9*, 1–15.
63. Corrado, S., & Sala, S. (2018). Food waste accounting along global and European food supply chains: State of the art and outlook. *Waste Manag.*, *79*, 120–s131.
64. Cosgrove, D. J. (2023). Structure and growth of plant cell walls. *Nat. Rev. Mol. Cell Biol., 25(5), 340–358.*
65. Counts, J. A., Zeldes, B. M., Lee, L. L., Straub, C. T., Adams, M. W. W., & Kelly, R. M. (2017). Physiological, metabolic and biotechnological features of extremely thermophilic microorganisms. *Wiley Interdiscip. Rev. Syst. Biol. Med.*, *9*, e1377.
66. Crippa, M., Solazzo, E., Guizzardi, D., Monforti-Ferrario, F., Tubiello, F. N., & Leip, A. (2021). Food systems are responsible for a third of global anthropogenic GHG emissions. *Nat. Food*, *2*, 198–209.
67. Curie, C., & Briat, J. F. (2003). Iron transport and signaling in plants. *Annu. Rev. Plant Biol.*, *54*(1), 183–206.
68. Darmon, N., & Drewnowski, A. (2015). Contribution of food prices and diet cost to socioeconomic disparities in diet quality and health: A systematic review and analysis. *Nutr. Rev.*, *73*, 643–660.
69. Darnhofer, I., Bellon, S., Dedieu, B., & Milestad, R. (2010). Adaptiveness to enhance the sustainability of farming systems: A review. *Agronomy for Sustainable Development*, *30*, 545–555.
70. Daszkiewicz, T. (2022). Food production in the context of global developmental challenges. *Agriculture*, *12*(6), 832.
71. Davis, D. R., Epp, M. D., & Riordan, H. D. (2004). Changes in USDA

food composition data for 43 garden crops, 1950 to 1999. *J. Am. Coll. Nutr.*, *23*(6), 669–682.

72. Day, K. J., Hutchings, M. J., & John, E. A. (2003). The effects of spatial pattern of nutrient supply on the early stages of growth in plant populations. *J. Ecol.*, *91*, 305–315.
73. de Kroon, H., Hendriks, M., van Ruijven, J., Ravenek, J., Padilla, F. M., Jongejans, E., Visser, E. J. W., & Mommer, L. (2012). Root responses to nutrients and soil biota: Drivers of species coexistence and ecosystem productivity. *J. Ecol.*, *100*, 6–15.
74. De Moura, F. F., Miloff, A., & Boy, E. (2015). Retention of provitamin A carotenoids in staple crops targeted for biofortification in Africa: Cassava, maize and sweet potato. *Crit. Rev. Food Sci. Nutr.*, *55*(9), 1246–1269.
75. De Vries, W., et al. (2022). Impacts of nutrients and heavy metals in European agriculture—current and critical inputs in relation to air, soil and water quality. *ETC-DI Report 2022/01*. Available from: https://www.eionet.europa.eu/etcs/etc-di/products/impacts-of-nutrients-and-heavy-metals-in-european-agriculture-current-and-critical-inputs-in-relation-to-air-soil-and-water-quality (accessed on 17 October 2024).
76. Del Carmen, O. M. M., Macías-Rodríguez, L. I., Santoyo, G., Farías-Rodríguez, R., & Valencia-Cantero, E. (2013a). *Medicago truncatula* increases its iron-uptake mechanisms in response to volatile organic compounds produced by *Sinorhizobium meliloti*. *Folia Microbiol. (Praha)*, *58*(6), 579–585.
77. Del Carmen, O. M. M., Velázquez-Becerra, C., Macías-Rodríguez, L. I., Santoyo, G., Flores-Cortez, I., Alfaro-Cuevas, R., et al. (2013b). *Arthrobacter agilis* UMCV2 induces iron acquisition in *Medicago truncatula* (strategy I plant) *in vitro* via dimethylhexadecylamine emission. *Plant Soil*, *362*(1), 51–66.
78. DeLonge, M. S., Miles, A., & Carlisle, L. (2016). Investing in the transition to sustainable agriculture. *Environmental Science & Policy*, *55*, 266–273.
79. Díaz, S., et al. (2018). Assessing nature's contributions to people: Recognizing culture, and diverse sources of knowledge, can improve assessments. *Science*, *359*, 270–272.
80. Dioula, B. M., Deret, H., Morel, J., et al. (2013). Enhancing the role of smallholder farmers in achieving sustainable food and nutrition security. *ICN 2 Second International Conference on Nutrition*, supported by the Food and Agricultural Organization of the United Nations and World Health Organization.
81. Dixon, R. A., & Postgate, J. R. (1971). Transfer of nitrogen-fixation genes by conjugation in *Klebsiella pneumoniae*. *Nature*, *234*(5323), 47–48.

82. Dlamini, S. P., Akanmu, A. O., & Babalola, O. O. (2022). Rhizospheric microorganisms: The gateway to sustainable plant health. *Front. Sustain. Food Syst.*, *6*, 925802.

83. D'Odorico, P., Carr, J. A., Davis, K. F., Dell'Angelo, J., & Seekell, D. A. (2019). Food inequality, injustice, and rights. *Bioscience*, *69*, 180–190.

84. Dong, B. C., Alpert, P., Zhang, Q., & Yu, F. H. (2015). Clonal integration in homogeneous environments increases performance in *Alternanthera philoxeroides*. *Oecologia*, *179*, 393–400.

85. Du, E., Terrer, C., Pellegrini, A. F. A., Ahlström, A., van Lissa, C. J., Zhao, X., Xia, N., Wu, X., & Jackson, R. B. (2020). Global patterns of terrestrial nitrogen and phosphorus limitation. *Nature Geoscience*, *13*, 221–226.

86. Du, J., Yu, F. H., Alpert, P., & Dong, M. (2009). Arbuscular mycorrhizal fungi reduce effects of physiological integration in *Trifolium repens*. *Annals of Botany*, *104*, 335–344.

87. Eichhorn, E., Van Der Ploeg, J. R., & Leisinger, T. (1999). Characterization of a two-component alkanesulfonate monooxygenase from *Escherichia coli*. *Journal of Biological Chemistry*, *274*(38), 26639–26646. https://doi.org/10.1074/jbc.274.38.26639.

88. Eswaran, H., Reich, P. F., & Padmanabhan, E. (2012). World soil resources opportunities and challenges. In B. Lal & R. Stewart (Eds.), *World Soil Resources and Food Security* (pp. 29–52). Boca Raton, FL: CRC-Press.

89. European Commission (2020). Italian government adopts targeted regularisation for migrant workers. Available from: ec.europa.eu/migrant-integration/news/italiangovernment-adopts-targeted-regularisation-for-migrantworkers. https://migrant-integration.ec.europa.eu/news /italian-government-adopts-targeted-regularisation-migrant-workers_en

90. European Commission (2021). Communication from the Commission to the European Parliament, the Council, the European Economic and Social Committee and the Committee of the Regions 'EU soil strategy for 2030–Reaping the benefits of healthy soils for people, food, nature and climate' (COM(2021) 699 final of 17 November 2021). https://eur-lex.europa.eu/legal-content/EN/TXT/PDF/?uri=CELEX:52021DC0699&from=EN (accessed on 17 October 2024).

91. Fan, M. S., Zhao, F. J., Fairweather-Tait, S. J., Poulton, P. R., Dunham, S. J., & McGrath, S. P. (2008). Evidence of decreasing mineral density in wheat grain over the last 160 years. *Journal of Trace Elements in Medicine and Biology*, *22*(4), 315–324. https://doi.org/10.1016/j.jtemb.2008.07.002.

92. FAO (Food and Agriculture Organization). (2009). *Global Agriculture Towards 2050, How to Feed the World 2050*. High level expert forum, Rome.

93. FAO (Food and Agriculture Organization). (2015). Healthy soils are the basis for healthy food production. Available from: https://www.fao.org/3/i4405e/i4405e.pdf (accessed on 17 October 2024).

94. FAO. (2020). Cereal supply and demand balances for sub-Saharan African countries. Rome, Italy: Food and Agriculture Organization of the United Nations.

95. FAO. (2020). *The State of Agricultural Commodity Markets 2020. Agricultural Markets and Sustainable Development: Global Value Chains, Smallholder Farmers and Digital Innovations*. Rome, Italy: FAO.

96. FAOSTAT. (2021). *Food and Agriculture Data*. Available from: http://www.fao.org/faostat/en/#home (accessed on 17 October 2024).

97. Farag, M. A., Ryu, C. M., Sumner, L. W., & Paré, P. W. (2006). GC–MS SPME profiling of rhizobacterial volatiles reveals prospective inducers of growth promotion and induced systemic resistance in plants. *Phytochemistry*, *67*(20), 2262–2268. https://doi.org/10.1016/j.phytochem.2006.07.021.

98. Farag, M. A., Zhang, H., & Ryu, C. M. (2013). Dynamic chemical communication between plants and bacteria through airborne signals: induced resistance by bacterial volatiles. *Journal of Chemical Ecology*, *39*(7), 1007–1018. https://doi.org/10.1007/s10886-013-0317-9.

99. Ferguson, B. J., et al. (2010). Molecular analysis of legume nodule development and autoregulation. *Journal of Integrative Plant Biology*, *52*, 61–76.

100. Ferrol, N., Tamayo, E., & Vargas, P. (2016). The heavy metal paradox in arbuscular mycorrhizas: from mechanisms to biotechnological applications. *Journal of Experimental Botany*, *67*(22), erw403. https://doi.org/10.1093/jxb/erw403.

101. Finkelstein, J. L., Mehta, S., Udipi, S. A., Ghugre, P. S., Luna, S. V., Wenger, M. J., Murray-Kolb, L. E., Przybyszewski, E. M., & Haas, J. D. (2015). A randomized trial of iron-biofortified pearl millet in school children in India. *Journal of Nutrition*, *145*(7), 1576–1581.

102. Foley, J. A., et al. (2011). Solutions for a cultivated planet. *Nature*, *478*, 337–342.

103. Food and Agriculture Organization of the United Nations. *FAOSTAT Statistical Database*. Rome: FAO; 2022.

104. Fore, H. H., Dongyu, Q., Beasley, D. M., & Ghebreyesus, T. A. (2020). Child malnutrition and COVID-19: the time to act is now. *The Lancet*, *6736/20*, 20–21. https://doi.org/10.1016/s0140-6736(20)31648-2.

105. Fransen, B., de Kroon, H., & Berendse, F. (1998). Root morphological

plasticity and nutrient acquisition of perennial grass species from habitats of different nutrient availability. *Oecologia*, *115*, 351–358.

106. Freitas, M. A., Medeiros, F. H., Carvalho, S. P., Guilherme, L. R., Teixeira, W. D., Zhang, H., & Paré, P. W. (2015). Augmenting iron accumulation in cassava by the beneficial soil bacterium *Bacillus subtilis* (GBO3). *Frontiers in Plant Science*, *6*, 596. https://doi.org/10.3389/fpls.2015.00596.
107. Gahan, J., & Schmalenberger, A. (2015). Arbuscular mycorrhizal hyphae in grassland select for a diverse and abundant hyphospheric bacterial community involved in sulfonate desulfurization. *Applied Soil Ecology*, *89*, 113–121. https://doi.org/10.1016/j.apsoil.2014.12.008.
108. Gannon, B., Kaliwile, C., Arscott, S. A., Schmaelzle, S., Chileshe, J., Kalungwana, N., et al. (2014). Biofortified orange maize is as efficacious as a vitamin A supplement in Zambian children even in the presence of high liver reserves of vitamin A: a community-based, randomized placebo-controlled trial. *American Journal of Clinical Nutrition*, *100*(6), 1541–1550.
109. García-Díaz, A., Bienes, R., Sastre, B., Novara, A., Gristina, L., & Cerdà, A. (2017). Nitrogen losses in vineyards under different types of soil groundcover. A field runoff simulator approach in central Spain. *Agriculture, Ecosystems & Environment*, *236*, 256–267. https://doi.org/10.1016/j.agee.2016.12.013.
110. Georgiou, K., Abramoff, R. Z., Harte, J., Riley, W. J., & Torn, M. S. (2017). Microbial community-level regulation explains soil carbon responses to long-term litter manipulations. *Nature Communications*, *8*(1), 1223.
111. Gerten, D., Heck, V., Jalava, M., Khan, M. S., Bodirsky, B. L., Fetzer, I., & Schellnhuber, H. J. (2020). Feeding ten billion people is possible within four terrestrial planetary boundaries. *Nature Sustainability*, *3*(3), 200–208.
112. Gezie, M., & Tejada, M. M. (2019). Farmer's response to climate change and variability in Ethiopia: A review. *Cogent Food & Agriculture*, *5*(1), 161.
113. GHI– Global Hunger Index. (2023). *Global Hunger Index 2023*. Retrieved from: www.globalhungerindex.org (accessed on 17 October 2024).
114. Ghosh, S., Bal, B., & Das, A. (2018). Enhancing manganese recovery from low-grade ores by using mixed culture of indigenously isolated bacterial strains. *Geomicrobiology Journal*, *35*(3), 242–246.
115. Giller, K. E., Hijbeek, R., Andersson, J. A., & Sumberg, J. (2021). Regenerative agriculture: An agronomic perspective. *Outlook on Agriculture*, *50*(1), 13–25.

116. Giovannetti, M., Tolosano, M., Volpe, V., Kopriva, S., & Bonfante, P. (2014). Identification and functional characterization of a sulfate transporter induced by both sulfur starvation and mycorrhiza formation in *Lotus japonicus*. *New Phytologist*, *204*(3), 609–619.

117. Govers, G., Merckx, R., Van Wesemael, B., & Van Oost, K. (2016). Soil conservation in the 21st century: Why we need smart intensification. *SOIL Discussions*, *3*, 45–59.

118. Govindarajulu, M., Pfeffer, P. E., Jin, H., Abubaker, J., Douds, D. D., Allen, J. W., & Shachar-Hill, Y. (2005). Nitrogen transfer in the arbuscular mycorrhizal symbiosis. *Nature*, *435*(7043), 819–823.

119. Grossman, R. B. (1983). Chapter 2: Entisols. In L. P. Wilding, N. E. Smeck, & G. F. Hall (Eds.), *Pedogenesis and Soil Taxonomy* (pp. 55–90). Amsterdam, The Netherlands: Elsevier.

120. Grote, U., Craswell, E., & Vlek, P. (2005). Nutrient flows in international trade: Ecology and policy issues. *Environmental Science & Policy*, *8*(5), 439–451.

121. Guerinot, M. L., Meidl, E. J., & Plessner, O. (1990). Citrate as a siderophore in *Bradyrhizobium japonicum*. *Journal of Bacteriology*, *172*(6), 3298–3303.

122. Guerrero-Galán, C., Garcia, K., Houdinet, G., & Zimmermann, S. D. (2018). Hc TOK1 participates in the maintenance of K+ homeostasis in the ectomycorrhizal fungus *Hebeloma cylindrosporum*, which is essential for the symbiotic K+ nutrition of *Pinus pinaster*. *Plant Signaling & Behavior*, *13*(6), e1480845.

123. Gunders, D. (2012). *Wasted: How America is Losing up to 40 Percent of its Food from Farm to Fork to Landfill*. Natural Resources Defense Council, Issue Paper 12-06-B.

124. Gustavsson, J., Cederberg, C., Sonesson, U., Otterdijk, R., & Mcybeck, A. (2011). *Global Food Losses and Food Waste: Extent, Causes and Prevention*. Food and Agriculture Organization of the United Nations.

125. Haas, J. D., Beard, J. L., Murray-Kolb, L. E., Del Mundo, A. M., Felix, A., & Gregorio, G. B. (2005). Iron-biofortified rice improves the iron stores of nonanemic Filipino women. *Journal of Nutrition*, *135*(12), 2823–2830.

126. Haas, J. D., Luna, S. V., Lung'aho, M. G., Wenger, M. J., Murray-Kolb, L. E., Beebe, S., & Egli, I. M. (2016). Consuming iron biofortified beans increases iron status in Rwandan women after 128 days in a randomized controlled feeding trial. *Journal of Nutrition*, *146*(8), 1586–1592.

127. Hadi, J., & Brightwell, G. (2021). Safety of alternative proteins: Technological, environmental and regulatory aspects of cultured meat,

plant-based meat, insect protein, and single-cell protein. *Foods*, *10*(6), 1226.

128. Hailu, K., Gashu, D., Joy, E. J. M., Alonso, S., Gizaw, S., & Gameda, S. (2022). Selenium concentration in cattle serum and fodder from two areas in Ethiopia with contrasting human selenium concentration. *Frontiers in Bioscience*, *27*, 200.
129. Hall, D., Guo, J., Dore, M., & Chow, C. C. (2009). The progressive increase of food waste in America and its environmental impact. *PLoS ONE*, *4*(11), e7940.
130. Han, J., Shi, J., Zeng, L., Xu, J., & Wu, L. (2015). Effects of nitrogen fertilization on the acidity and salinity of greenhouse soils. *Environmental Science and Pollution Research*, *22*(4), 2976–2986.
131. Hargrove, W. L. (2003). Soil fertility evaluation by plant analysis. *Soil Science Society of America Journal*, *47*(4), 189–193.
132. Harris, J. (2009). Soil microbial communities and restoration ecology: Facilitators or followers? *Science*, *325*(5940), 573–574.
133. Harrison, M. J., Dewbre, G. R., & Liu, J. (2002). A phosphate transporter from *Medicago truncatula* involved in the acquisition of phosphate released by arbuscular mycorrhizal fungi. *The Plant Cell*, *14*(10), 2413–2429.
134. Hatfield, J. L., Sauer, T. J., & Cruse, R. M. (2017). Chapter one: Soil—The forgotten piece of the water, food, energy nexus. In D. L. Sparks (Ed.), *Advances in Agronomy* (pp. 1–46). New York, NY: Academic Press.
135. HBM4EU. (2022). *Cadmium* (Policy Brief). European Human Biomonitoring Initiative. Retrieved from:–https://www.hbm4eu.eu/wp-content/uploads/2022/07/HBM4EU_Policy-Brief-Cadmium.pdf. https://www.hbm4eu.eu/policy_briefs/cadmium/
136. Hengl, T., de Jesus, J. M., Heuvelink, G. B., Ruiperez Gonzalez, M., Kilibarda, M., Blagotic, A., & Kempen, B. (2017). SoilGrids250 m: Global gridded soil information based on machine learning. *PLoS ONE*, *12*(2), e0169748.
137. Hermans, C., Vuylsteke, M., Coppens, F., Cristescu, S. M., Harren, F. J., Inzé, D., et al. (2010). Systems analysis of the responses to long-term magnesium deficiency and restoration in *Arabidopsis thaliana*. *New Phytologist, 187*(1), 132–144.
138. Hernández-Calderón, E., Aviles-Garcia, M. E., Castulo-Rubio, D. Y., Macías-Rodríguez, L., Ramírez, V. M., Santoyo, G., López-Bucio, J., & Valencia-Cantero, E. (2018). Volatile compounds from beneficial or pathogenic bacteria differentially regulate root exudation, transcription

of iron transporters, and defense signaling pathways in *Sorghum bicolor*. *Plant Molecular Biology, 96*(3), 291–304.

139. Hestrin, R., Hammer, E. C., Mueller, C. W., & Lehmann, J. (2019). Synergies between mycorrhizal fungi and soil microbial communities increase plant nitrogen acquisition. *Communications Biology, 2*, 233.
140. Hodge, A. (2004). The plastic plant: Root responses to heterogeneous supplies of nutrients. *New Phytologist, 162*, 9–24.
141. Hodge, A., Robinson, D., & Fitter, A. H. (2000). An arbuscular mycorrhizal inoculum enhances root proliferation in, but not nitrogen capture from, nutrient-rich patches in soil. *New Phytologist, 145*, 575–584.
142. Horton, R. E. (1933). The role of infiltration in the hydrologic cycle. *Eos, Transactions American Geophysical Union, 14*, 446–460.
143. Hotz, C., Loechl, C., de Brauw, A., Eozenou, P., Gilligan, D., Moursi, M., et al. (2012). A large-scale intervention to introduce orange sweet potato in rural Mozambique increases vitamin A intakes among children and women. *British Journal of Nutrition, 108*(1), 163–176.
144. Hotz, C., Loechl, C., Lubowa, A., Tumwine, J. K., Ndeezi, G., Nandutu Masawi, A., et al. (2012b). Introduction of β-carotene-rich orange sweet potato in rural Uganda resulted in increased vitamin A intakes among children and women and improved vitamin A status among children. *Journal of Nutrition, 142*(10), 1871–1880.
145. Hu, W., Ran, J., Dong, L., Du, Q., Ji, M., Yao, S., Sun, Y., Gong, C., Hou, Q., Gong, H., et al. (2021). Aridity-driven shift in biodiversity–soil multifunctionality relationships. *Nature Communications, 12*, 5350.
146. Hutchings, M. J., & de Kroon, H. D. (1994). Foraging in plants: The role of morphological plasticity in resource acquisition. *Advances in Ecological Research, 25*, 159–238.
147. Hutchings, M., & Wijesinghe, D. (2008). Performance of a clonal species in patchy environments: Effects of environmental context on yield at local and whole-plant scales. *Evolutionary Ecology, 22*, 313–324.
148. International Fertilizer Association. (2022). Paris, France.
149. International Labor Organization. (2020). *ILO Monitor: COVID-19 and the world of work. Third edition*. Geneva. Retrieved from: www.ilo.org/global/topics/coronavirus/impacts-and-responses/WCMS_743146/lang--en/index.htm (accessed on 17 October 2024).
150. Islam, S. (2019). Conservation agriculture based sustainable intensification: Increasing yields and water productivity for smallholders of the Eastern Gangetic Plains. *Field Crops Research, 17*(1), 25–42.

151. ISRIC–World Soil Information. (2020). SoilGrids. Available from: https://www.soilgrids.org/ (accessed on 17 October 2024).
152. Jackson, R. B., & Caldwell, M. M. (1993). The scale of nutrient heterogeneity around individual plants and its quantification with geostatistics. *Ecology, 74*, 612–614.
153. Jacoby, R., Peukert, M., Succurro, A., Koprivova, A., & Kopriva, S. (2017). The role of soil microorganisms in plant mineral nutrition—Current knowledge and future directions. *Frontiers in Plant Science, 8*, 1617.
154. Jamil, K. M., Brown, K. H., Jamil, M., Peerson, J. M., Keenan, A. H., Newman, J. W., & Haskell, M. J. (2012). Daily consumption of orange-fleshed sweet potato for 60 days increased plasma β-carotene concentration but did not increase total body vitamin A pool size in Bangladeshi women. *Journal of Nutrition, 142*(10), 1896–1902.
155. Jarstfer, A., Farmer-Koppenol, P., & Sylvia, D. M. (1998). Tissue magnesium and calcium affect arbuscular mycorrhiza development and fungal reproduction. *Mycorrhiza, 7*(5), 237–242.
156. Jayedi, A., Soltani, S., Abdolshahi, A., & Shab-Bidar, S. (2020). Healthy and unhealthy dietary patterns and the risk of chronic disease: An umbrella review of meta-analyses of prospective cohort studies. *British Journal of Nutrition, 124*, 1133–1144.
157. Jensen, E. S., Bedoussac, L., Carlsson, G., Journet, E. P., Justes, E., & Hauggaard-Nielsen, H. (2015). Enhancing yields in organic crop production by eco-functional intensification. *Sustainable Agriculture Research, 4*(3), 42–50.
158. Jiang, F., Zhang, L., Zhou, J., George, T. S., & Feng, G. (2021). Arbuscular mycorrhizal fungi enhance mineralisation of organic phosphorus by carrying bacteria along their extraradical hyphae. *New Phytologist, 230*(1), 304–315.
159. Jiang, Y., Zhou, H., Chen, L., Yuan, Y., Fang, H., Luan, L., Chen, Y., Wang, X., Liu, M., Li, H., et al. (2018). Nematodes and microorganisms interactively stimulate soil organic carbon turnover in the macroaggregates. *Frontiers in Microbiology, 9*, 2803.
160. Jing, X., Cui, Q., Li, X., Yin, J., Ravichandran, V., Pan, D., Fu, J., Tu, Q., Wang, H., Bian, X., & Zhang, Y. (2020). Engineering *Pseudomonas protegens* Pf-5 to improve its antifungal activity and nitrogen fixation. *Microbial Biotechnology, 13*(1), 118–133.
161. Jobson, E. M., Johnston, R. E., Oiestad, A. J., Martin, J. M., & Giroux, M. J. (2019). The impact of the wheat Rht-B1b semi-dwarfing allele on

photosynthesis and seed development under field conditions. *Frontiers in Plant Science, 10*, 51.

162. Johnson, D., IJdo, M., Genney, D. R., Anderson, I. C., & Alexander, I. J. (2005). How do plants regulate the function, community structure, and diversity of mycorrhizal fungi? *Journal of Experimental Botany*, *56*, 1751–1760.

163. Johnson, N. C. (2010). Resource stoichiometry elucidates the structure and function of arbuscular mycorrhizas across scales. *New Phytologist*, *185*, 631–647.

164. Jones, D. L. (2016). Soil microbial biomass and nutrient cycling relationships. *Soil Biology and Biochemistry*, *102*, 13–14.

165. Kaiser, B. N., Gridley, K. L., Ngaire Brady, J., Phillips, T., & Tyerman, S. D. (2005). The role of molybdenum in agricultural plant production. *Annals of Botany*, *96*(5), 745–754.

166. Kakaei, H., Nourmoradi, H., Bakhtiyari, S., et al. (2022). Effect of COVID-19 on food security, hunger, and food crisis. In *COVID-19 and the Sustainable Development Goals* (Chapter One).

167. Kamran, S., Shahid, I., Baig, D. N., Rizwan, M., Malik, K. A., & Mehnaz, S. (2017). Contribution of zinc solubilizing bacteria in growth promotion and zinc content of wheat. *Frontiers in Microbiology*, *8*, 2593.

168. Kanchiswamy, C. N., Malnoy, M., & Maffei, M. E. (2015). Chemical diversity of microbial volatiles and their potential for plant growth and productivity. *Frontiers in Plant Science*, *6*, 151.

169. Kane, D. A., Bradford, M. A., Fuller, E., Oldfield, E. E., & Wood, S. A. (2021). Soil organic matter protects US maize yields and lowers crop insurance payouts under drought. *Environmental Research Letters*, *16*(4), 044018.

170. Karandashov, V., & Bucher, M. (2005). Symbiotic phosphate transport in arbuscular mycorrhizas. *Trends in Plant Science*, *10*, 22–29.

171. Kawaguchi, M., & Minamisawa, K. (2010). Plant–microbe communications for symbiosis. *Plant Cell Physiology*, *51*, 1377–1380.

172. Keating, B. A., Herrero, M., Carberry, P. S., Gardner, J., & Cole, M. B. (2014). Food wedges: framing the global food demand and supply challenge towards 2050. *Global Food Security*, *3*, 125–132.

173. Kent, G. (2008). The benefits of world hunger. *UN Chronicle*, No. 2/3, 81–82.

174. Kertesz, M. A. (2000). Riding the sulfur cycle–metabolism of sulfonates and sulfate esters in gram-negative bacteria. *FEMS Microbiology Reviews*, *24*(2), 135–175.

175. Kertesz, M. A., & Mirleau, P. (2004). The role of soil microbes in plant sulfur nutrition. *Journal of Experimental Botany*, *55*(404), 1939–1945.
176. Kertesz, M. A., Fellows, E., & Schmalenberger, A. (2007). Rhizobacteria and plant sulfur supply. *Advances in Applied Microbiology*, *62*, 235–268.
177. King, J., & Quinn, J. (1997). The utilization of organosulphonates by soil and freshwater bacteria. *Letters in Applied Microbiology*, *24*(6), 474–478.
178. Kiri, I. (2023). Mechanisms of nutrient uptake and assimilation processes in some plants: A review. *Dutse Journal of Pure and Applied Sciences*, *9*, 223–237.
179. Kloepper, J. W., Leong, J., Teintze, M., & Schroth, M. N. (1980). Enhanced plant growth by siderophores produced by plant growth-promoting rhizobacteria. *Nature*, *286*(5776), 885–886.
180. Kochhar, N., Shrivastava, S., Ghosh, A., Rawat, V. S., Sodhi, K. K., & Kumar, M. (2022). Perspectives on the microorganism of extreme environments and their applications. *Current Research in Microbial Sciences*, *3*, 100134.
181. Kong, W. L., Wang, Y. H., & Wu, X. Q. (2021). Enhanced iron uptake in plants by volatile emissions of *Rahnella aquatilis* JZ-GX1. *Frontiers in Plant Science*, *12*, 704000.
182. Kopittke, P. M., Menzies, N. W., Wang, P., McKenna, B. A., & Lombi, E. (2019). Soil and the
183. intensification of agriculture for global food security. *Environment international*, *132*, 105078
184.
185. Kothari, S., Marschner, H., & Römheld, V. (1991). Effect of a vesicular–arbuscular mycorrhizal fungus and rhizosphere micro-organisms on manganese reduction in the rhizosphere and manganese concentrations in maize (*Zea mays* L.). *New Phytologist*, *117*(4), 649–655.
186. Kotir, J. H. (2011). Climate change and variability in Sub-Saharan Africa: A review of current and future trends and impacts on agriculture and food security. *Environment, Development and Sustainability*, *13*, 587–605.
187. Krithika, S., & Balachandar, D. (2016). Expression of zinc transporter genes in rice as influenced by zinc-solubilizing *Enterobacter cloacae* strain ZSB14. *Frontiers in Plant Science*, *7*, 446.
188. Kumar, S., Joshi, P. K., & Pal, S. (2004). Impact of vegetable research in India. *National Centre for Agricultural Economics and Policy Research*, Pusa, New Delhi. Retrieved from: www.ncap.res.in/upload_files/workshop/wsp13.pdf. https://krishi.icar.gov.in/jspui/bitstream/123456789/779/1/wsp13.pdf

189. Kurothe, R., Kumar, G., Singh, R., Singh, H., Tiwari, S., Vishwakarma, A., et al. (2014). Effect of tillage and cropping systems on runoff, soil loss and crop yields under semiarid rainfed agriculture in India. *Soil & Tillage Research*, *140*, 126–134.

190. La Frano, M. R., de Moura, F. F., Boy, E., Lönnerdal, B., & Burri, B. J. (2014). Bioavailability of iron, zinc, and provitamin A carotenoids in biofortified staple crops. *Nutrition Reviews*, *72*(5), 289–307.

191. La, J., Roberts, N. H., & Yafi, F. A. (2018). Diet and men's sexual health. *Sexual Medicine Reviews*, *6*, 54–68.

192. Lal, R. (2003). Soil erosion and the global need for sustainable agriculture. *Soil and Tillage Research*, *1*(1), 23–32.

193. Lal, R. (2020). Soil organic matter and water retention. *Agronomy Journal*, *112*(5), 3265–3277.

194. Landrigan, P. J., et al. (2018). The Lancet Commission on pollution and health. *The Lancet*, *391*, 462–512.

195. Lee, Y. J., & George, E. (2005). Contribution of mycorrhizal hyphae to the uptake of metal cations by cucumber plants at two levels of phosphorus supply. *Plant and Soil, 278*(1), 361–370.

196. Leff, J. W., Bardgett, R. D., Wilkinson, A., Jackson, B. G., Pritchard, W. J., De Long, J. R., Oakley, S., Mason, K. E., Ostle, N. J., Johnson, D., et al. (2018). Predicting the structure of soil communities from plant community taxonomy, phylogeny, and traits. *ISME Journal, 12*, 1794–1805.

197. Lehmann, J., & Rondon, M. (2006). Biochar soil management on highly weathered soils in the humid tropics. In *Biological Approaches to Sustainable Soil Systems* (pp. 517–530). CRC Press.

198. Levasseur, P., et al. (2021). The health and socioeconomic costs of exposure to soil pollution: Evidence from three polluted mining and industrial sites in Europe. *Journal of Public Health, 30*, 2533–2546.

199. Li, G., et al. (2018). Urban soil and human health: A review. *European Journal of Soil Science, 69*, 196–215.

200. Li, X. L., Marschner, H., & George, E. (1991). Acquisition of phosphorus and copper by VA-mycorrhizal hyphae and root-to-shoot transport in white clover. *Plant and Soil, 136*(1), 49–57.

201. Li, X., & Christie, P. (2001). Changes in soil solution Zn and pH and uptake of Zn by arbuscular mycorrhizal red clover in Zn-contaminated soil. *Chemosphere, 42*(2), 201–207.

202. Lim, B. L., Yeung, P., Cheng, C., & Hill, J. E. (2007). Distribution and diversity of phytate-mineralizing bacteria. *ISME Journal, 1*(4), 321–330.

203. Limpens, E., & Bisseling, T. (2003). Signaling in symbiosis. *Current Opinion in Plant Biology, 6*, 343–350.

204. Ling, N., Wang, T., & Kuzyakov, Y. (2022). Rhizosphere bacteriome structure and functions. *Nature Communications, 13*, 836.

205. Liu, C., Gong, X., Dang, K., Li, J., Yang, P., Gao, X., Deng, X., & Feng, B. (2020). Linkages between nutrient ratio and the microbial community in rhizosphere soil following fertilizer management. *Environmental Research, 184*, 109261.

206. Liu, H., Li, S., Qiang, R., Lu, E., Li, C., Zhang, J., & Gao, Q. (2022). Response of soil microbial community structure to phosphate fertilizer reduction and combinations of microbial fertilizer. *Frontiers in Environmental Science, 10*, 899727.

207. Liu, X. M., & Zhang, H. (2015). The effects of bacterial volatile emissions on plant abiotic stress tolerance. *Frontiers in Plant Science, 6*, 774.

208. Lopez-Bucio, J., et al. (2003). The role of nutrient availability in regulating root architecture. *Current Opinion in Plant Biology, 6*, 280–287.

209. Louhichi, K., Riesgo, L., & Paloma, S. G. (2020). The role of smallholder farms in food and nutrition security. *Food and Nutrition Security, 251.*

210. Low, J. W., Arimond, M., Osman, N., Cunguara, B., Zano, F., & Tschirley, D. (2007). A food-based approach introducing orange-fleshed sweet potatoes increased vitamin A intake and serum retinol concentrations in young children in rural Mozambique. *Journal of Nutrition, 137*, 1320–1327.

211. Loydi, A., Lohse, K., Otte, A., Donath, T. W., & Eckstein, R. L. (2013). Distribution and effects of tree leaf litter on vegetation composition and biomass in a forest–grassland ecotone. *Journal of Plant Ecology, 7*, 264–275.

212. Lugtenberg, B., & Kamilova, F. (2009). Plant-growth-promoting rhizobacteria. *Annual Review of Microbiology, 63*, 541–556.

213. Maddela, N. R., et al. (2022). Major contaminants of emerging concern in soils: A perspective on potential health risks. *RSC Advances, 12*, 12396–12415.

214. Maggio, A., Van Criekinge, T., & Malingreau, J. P. (2015). Global food security 2030: Assessing trends with a view to guiding future EU policies. *JRC Science and Policy Reports*. Brussels, Belgium: European Commission DG Joint Research Centre.

215. Marastoni, L., Pii, Y., Maver, M., Valentinuzzi, F., Cesco, S., & Mimmo, T. (2019). Role of *Azospirillum brasilense* in triggering different Fe chelate reductase enzymes in cucumber plants subjected to both nutrient

deficiency and toxicity. *Plant Physiology and Biochemistry, 136*, 118–126.

216. Marles, R. J. (2017). Mineral nutrient composition of vegetables, fruits and grains: The context of reports of apparent historical declines. *Journal of Food Composition and Analysis, 57*, 93–103.
217. Marschner, P. (2012). *Mineral Nutrition of Higher Plants*. Academic Press.
218. Martínez-Medina, A., Van Wees, S. C., & Pieterse, C. M. (2017). Airborne signals from *Trichoderma* fungi stimulate iron uptake responses in roots resulting in priming of jasmonic acid-dependent defenses in shoots of *Arabidopsis thaliana* and *Solanum lycopersicum*. *Plant Cell and Environment, 40*(11), 2691–2705.
219. Masood, S., & Bano, A. (2016). Mechanism of potassium solubilization in agricultural soils by the help of soil microorganisms. In *Potassium Solubilizing Microorganisms for Sustainable Agriculture* (pp. 137–147). Springer.
220. Matison, S. P., Mather, K. A., Flood, V. M., & Reppermund, S. (2021). Associations between nutrition and the incidence of depression in middle-aged and older adults: A systemic review and meta-analysis of prospective observational population-based studies. *Ageing Research Reviews, 70*, 101403.
221. Matthes, M. S., Robil, J. M., & Mcsteen, P. (2020). From element to development: The power of the essential micronutrient boron to shape morphological processes in plants. *Journal of Experimental Botany, 71*(5), 1681–1693.
222. Mayer, A., Hausfather, Z., Jones, A. D., & Silver, W. L. (2018). The potential of agricultural land management to contribute to lower global surface temperatures. *Science Advances, 4*, aaq0932.
223. Meena, V., Maurya, B., & Bahadur, I. (2014). Potassium solubilization by bacterial strain in waste mica. *Bangladesh Journal of Botany, 43*(2), 235–237.
224. Mekuria, W. (2018). The link between agricultural production and population dynamics in Ethiopia: A review. *Advances in Plants & Agriculture Research, 8*(2), 348–353.
225. Meldau, D. G., Meldau, S., Hoang, L. H., Underberg, S., Wünsche, H., & Baldwin, I. T. (2013). Dimethyl disulfide produced by the naturally associated bacterium *Bacillus* sp. B55 promotes *Nicotiana attenuata* growth by enhancing sulfur nutrition. *Plant Cell, 25*(7), 2731–2747.
226. Mendes, R., Garbeva, P., & Raaijmakers, J. M. (2013). The rhizosphere microbiome: Significance of plant beneficial, plant pathogenic, and human

pathogenic microorganisms. *FEMS Microbiology Reviews, 37*, 634–663.

227. Mercader, J. (2009). Mozambican grass seed consumption during the middle stone age. *Science, 326*, 1680–1683.

228. Mishra, D. B. B., & Nayak, D. K. C. (2004). Organic farming for sustainable agriculture. *Journal of Agriculture, 24*(2), 25.

229. Mitchell, D. B., Vogel, K., Weimann, B. J., Pasamontes, L., & Van Loon, A. P. (1997). The phytase subfamily of histidine acid phosphatases: Isolation of genes for two novel phytases from the fungi *Aspergillus terreus* and *Myceliophthora thermophila*. *Microbiology, 143*(1), 245–252.

230. Mitter, B., Pfaffenbichler, N., & Sessitsch, A. (2016). Plant–microbe partnerships in 2020. *Microbial Biotechnology, 9*, 635–640.

231. Miwa, K., Kamiya, T., & Fujiwara, T. (2009). Homeostasis of the structurally important micronutrients B and Si. *Current Opinion in Plant Biology, 12*(3), 307–311.

232. Mohajan, H. K. (2022). Food insecurity and malnutrition of Africa: A combined attempt can reduce them. *Journal of Economic Development, Environment and People, 11*(1).

233. Mohanty, S., Ghosh, S., Nayak, S., & Das, A. (2017). Isolation, identification and screening of manganese solubilizing fungi from low-grade manganese ore deposits. *Geomicrobiology Journal, 34*(4), 309–316.

234. Montanarella, L., & Panagos, P. (2021). The relevance of sustainable soil management within the European Green Deal. *Land Use Policy, 100*, 104950.

235. Montanarella, L., Yagi, K., Krasilnikov, P., Alavi Panah, S. K., Mendonça Santos, M., McKenzie, N., & Nachtergaele, F. (2015). The status of the world's soil resources. Rome, Italy: FAO & ITPS.

236. Montejano-Ramírez, V., García-Pineda, E., & Valencia-Cantero, E. (2020). Bacterial compound N, N-dimethylhexadecylamine modulates expression of iron deficiency and defense response genes in *Medicago truncatula* independently of the jasmonic acid pathway. *Plants, 9*(5), 624.

237. Morcillo, R. J. L., & Manzanera, M. (2021). The effects of plant-associated bacterial exopolysaccharides on plant abiotic stress tolerance. *Metabolites, 11*(6), 337.

238. Morrens, B., et al. (2012). Environmental justice under our skin? Socio-stratifying human biomonitoring results of adolescents living near an industrial hotspot in Flanders, Belgium. In *Looking Within: Finding an Environmental Justice and Global Citizenship Lens*. Brill Publishing, Netherlands.

239. Mueller, N. D., Gerber, J. S., Johnston, M., Ray, D. K., Ramankutty, N., & Foley, J. A. (2012). Closing yield gaps through nutrient and water management. *Nature, 490*, 254–257.

240. Muñoz López, M. d. M., Cecilia Revelo, M., & Pachón, H. (2008). El consumo y la producción familiar de fríjol, maíz, yuca, batata y arroz en un municipio rural en Colombia: Evaluación de la posibilidad de implementar la biofortificación de cultivos. *Perspectivas en Nutrición Humana, 10*, 11–21.

241. Mus, F., Crook, M. B., Garcia, K., Garcia Costas, A., Geddes, B. A., Kouri, E. D., Paramasivan, P., Ryu, M. H., Oldroyd, G. E. D., Poole, P. S., Udvardi, M. K., Voigt, C. A., Ané, J. M., & Peters, J. W. (2016). Symbiotic nitrogen fixation and the challenges to its extension to nonlegumes. *Applied and Environmental Microbiology, 82*(13), 3698–3710.

242. Mutyasira, V., Hoag, D., Pendell, D., & Yildiz, F. (2018). The adoption of sustainable agricultural practices by smallholder farmers in Ethiopian highlands: An integrative approach. *Cogent Food & Agriculture, 4*(1), 1552439.

243. Mwololo, H. M., Nzuma, J. M., & Githinji, L. M. (2022). Determinants of farmer empowerment in agriculture in Kenya: A Tobit approach. *Heliyon, 8*(12), e11888.

244. Nagothu, U. S. (2016). *Climate Change and Agricultural Development: Improving Resilience through Climate Smart Agriculture, Agroecology and Conservation*. Routledge.

245. Nagy, R., Karandashov, V., Chague, V., Kalinkevich, K., Tamasloukht, M. B., Xu, G., et al. (2005). The characterization of novel mycorrhiza-specific phosphate transporters from *Lycopersicon esculentum* and *Solanum tuberosum* uncovers functional redundancy in symbiotic phosphate transport in solanaceous species. *Plant Journal, 42*(2), 236–250.

246. Nannipieri, P., Giagnoni, L., Landi, L., & Renella, G. (2011). Role of phosphatase enzymes in soil. In E. K. Bunemann, A. Oberson, & E. Frossard (Eds.), *Phosphorus in Action* (pp. 215–243). Springer, Berlin, Heidelberg.

247. Neal, A. L., Hughes, D. J., Clark, I. M., Jansson, J. K., & Hirsch, P. R. (2021). Microbiome aggregated traits and assembly are more sensitive to soil management than diversity. *mSystems*, *6*(e01056–20).

248. Nehls, U. (2008). Mastering ectomycorrhizal symbiosis: The impact of carbohydrates. *Journal of Experimental Botany*, *59*, 1097–1108.

249. Nestel, P., Bouis, H. E., Meenakshi, J. V., & Pfeiffer, W. (2006). Biofortification of staple food crops. *Journal of Nutrition*, *136*, 1064–1067.

250. Nettleton, W. D., & Peterson, F. F. (1983). Aridisols. In L. P. Wilding, N. E. Smeck, & G. F. Hall (Eds.), *Pedogenesis and Soil Taxonomy* (pp. 165–215). Elsevier, Amsterdam, The Netherlands.

251. Ngoune, L. T., & Shelton, C. M. (2020). Factors affecting yields of crops. IntechOpen.

252. Nguyen, T. D., Cavagnaro, T. R., & Watts-Williams, S. J. (2019). The effects of soil phosphorus and zinc availability on plant responses to mycorrhizal fungi: A physiological and molecular assessment. *Scientific Reports*, *9*(1), 1–13.

253. Nielsen, U. N., Wall, D. H., & Six, J. (2015). Soil biodiversity and the environment. *Annual Review of Environment and Resources*, *40*, 63–90.

254. Nogueira, M. A., & Cardoso, E. J. B. N. (2003). Mycorrhizal effectiveness and manganese toxicity in soybean as affected by soil type and endophyte. *Scientia Agricola*, *60*, 329–335.

255. Nordin, S. M., & Nordin, K. J. (2017). Food the source of nutrition. *World Nutrition*, *8*(1), 87–89.

256. Obada, A. A., Msughter, A. E., & Namadi, H. M. (2021). Hyper prevalence of malnutrition in Nigeria context. *Biomedical Journal of Scientific and Technical Research*, *39*(1).

257. Oertel, C., Matschullat, J., Zurba, K., Zimmermann, F., & Erasmi, S. (2016). Greenhouse gas emissions from soils: A review. *Geochemistry*, *76*, 327–352.

258. Olanrewaju, O. S., & Babalola, O. O. (2022). The rhizosphere microbial complex in plant health: A review of interaction dynamics. *Journal of Integrative Agriculture*, *21*, 2168–2182.

259. Osabohien, R., Adeleye, N., & Alwis, T. D. (2020). Agro-financing and food production in Nigeria. *Heliyon*, *6*(e04001).

260. Ougier, E., et al. (2021). Burden of osteoporosis and costs associated with human biomonitored cadmium exposure in three European countries: France, Spain, and Belgium. *International Journal of Hygiene and Environmental Health*, *234*, 113747.

261. Pajares, S., & Bohannan, B. J. (2016). Ecology of nitrogen-fixing, nitrifying, and denitrifying microorganisms in tropical forest soils. *Frontiers in Microbiology*, *7*, 1045.

262. Palmer, A. C., Healy, K., Barffour, M. A., Siamusantu, W., Chileshe, J., Schulze, K. J., et al. (2016b). Provitamin A carotenoid-biofortified maize consumption increases pupillary responsiveness among Zambian children in a randomized controlled trial. *Journal of Nutrition*, *146*(12), 2551–2558.

263. Palmer, A. C., Siamusantu, W., Chileshe, J., Schulze, K. J., Barffour, M.,

Craft, N. E., et al. (2016). Provitamin A-biofortified maize increases serum β-carotene, but not retinol, in marginally nourished children: A cluster-randomized trial in rural Zambia. *American Journal of Clinical Nutrition*, *104*(1), 181–190.

264. Pandey, A., Tripathi, A., Srivastava, P., & Choudhary, K. K. (2019). Plant growth-promoting microorganisms in sustainable agriculture. In A. Kumar, A. K. Singh, & K. K. Choudhary (Eds.), *Role of Plant Growth Promoting Microorganisms in Sustainable Agriculture and Nanotechnology* (pp. 1–19). Woodhead Publishing.

265. Paszkowski, U., Kroken, S., Roux, C., & Briggs, S. P. (2002). Rice phosphate transporters include an evolutionarily divergent gene specifically activated in arbuscular mycorrhizal symbiosis. *Proceedings of the National Academy of Sciences of the USA*, *99*(20), 13324–13329.

266. Peng, L., Shan, X., Yang, Y., Wang, Y., Druzhinina, I. S., Pan, X., Jin, W., He, X., Wang, X., Zhang, X., Martin, F. M., & Yuan, Z. (2021). Facultative symbiosis with a saprotrophic soil fungus promotes potassium uptake in American sweetgum trees. *Plant Cell & Environment*, *44*(8), 2793–2809.

267. Pereira, R. V., Filgueiras, C. C., Dória, J., Peñaflor, M. F. G. V., & Willett, D. S. (2021). The effects of biostimulants on induced plant defense. *Frontiers in Agronomy*, *3*, 630596.

268. Perner, H., Schwarz, D., Bruns, C., Mäder, P., & George, E. (2007). Effect of arbuscular mycorrhizal colonization and two levels of compost supply on nutrient uptake and flowering of pelargonium plants. *Mycorrhiza*, *17*(5), 469–474.

269. Pfeiffer, W. H., & McClafferty, B. (2007). HarvestPlus: Breeding crops for better nutrition. *Crop Science*, *47*(S3), S88–S105.

270. Philippot, L., et al. (2013). The ecological significance of the diversity of soil microbial communities. *Soil Biology and Biochemistry*, *45*, 185–195.

271. Prasad, D., Verma, N., Bakshi, M., Narayan, O. P., Singh, A. K., Dua, M., & Johri, A. K. (2019). Functional characterization of a magnesium transporter of root endophytic fungus *Piriformospora indica*. *Frontiers in Microbiology*, *9*, 3231.

272. Preda, V. G., & Săndulescu, O. (2019). Communication is the key: Biofilms, quorum sensing, formation and prevention. *Discoveries*, *7*, e100.

273. Prity, S. A., Sajib, S. A., Das, U., Rahman, M. M., Haider, S. A., & Kabir, A. H. (2020). Arbuscular mycorrhizal fungi mitigate Fe deficiency symptoms in sorghum through phytosiderophore-mediated Fe mobilization and restoration of redox status. *Protoplasma*, *257*(5), 1373–1385.

274. Pulighe, G., & Lupia, F. (2020). Food first: COVID-19 outbreak and cities

lockdown a booster for a wider vision on urban agriculture. *Sustainability*, *12*(12), 5012.

275. Qian, Y. Q., Luo, D., Gong, G., Han, L., Ju, G. S., & Sun, Z. Y. (2014). Effects of spatial scale of soil heterogeneity on the growth of a clonal plant producing both spreading and clumping ramets. *Journal of Plant Growth Regulation*, *33*, 214–221.
276. Quiroga, G., Erice, G., Aroca, R., & Ruiz-Lozano, J. M. (2020). Elucidating the possible involvement of maize aquaporins in the plant boron transport and homeostasis mediated by *Rhizophagus irregularis* under drought stress conditions. *International Journal of Molecular Sciences*, *21*(5), 1748.
277. Rajkumar, M., Ae, N., Prasad, M. N. V., & Freitas, H. (2010). Potential of siderophore-producing bacteria for improving heavy metal phytoextraction. *Trends in Biotechnology*, *28*(3), 142–149.
278. Rakhra, V., Galappaththy, S. L., Bulchandani, S., & Cabandugama, P. K. (2020). Obesity and the western diet: How we got here. *Missouri Medicine*, *117*, 536–538.
279. Ramankutty, N., Evan, A. T., Monfreda, C., & Foley, J. A. (2008). Farming the planet: 1. Geographic distribution of global agricultural lands in the year 2000. *Global Biogeochemical Cycles*, *22*, 1–19.
280. Rausch, C., Daram, P., Brunner, S., Jansa, J., Laloi, M., Leggewie, G., Amrhein, N., & Bucher, M. (2001). A phosphate transporter expressed in arbuscule-containing cells in potato. *Nature*, *414*(6862), 462–465.
281. Read, D., & Perez-Moreno, J. (2003). Mycorrhizas and nutrient cycling in ecosystems—A journey towards relevance? *New Phytologist*, *157*, 475–492.
282. Reganold, J. P., et al. (2010). Organic agriculture in the twenty-first century. *Nature Plants*, *2*(2), 15221.
283. Reidsma, P., Wolf, J., Kanellopoulos, A., Schaap, B. F., Mandryk, M., Verhagen, J., et al. (2015). Climate change impact and adaptation research requires integrated assessment and farming systems analysis: A case study in the Netherlands. *Environmental Research Letters*, *10*(4), 045004.
284. Reynolds, H. L., & Haubensak, K. A. (2009). Soil fertility, heterogeneity, and microbes: Towards an integrated understanding of grassland structure and dynamics. *Applied Vegetation Science*, *12*, 33–44.
285. Riaz, N., & Guerinot, M. L. (2021). All together now: Regulation of the iron deficiency response. *Journal of Experimental Botany*, *72*(6), 2045–2055.
286. Richardson, A. E., & Simpson, R. J. (2011). Soil microorganisms

mediating phosphorus availability update on microbial phosphorus. *Plant Physiology*, *156*(3), 989–996.

287. Rinklebe, J., et al. (2019). Health risk assessment of potentially toxic elements in soils along the Central Elbe River, Germany. *Environment International*, *126*, 76–88.

288. Ritchie, H. (2017). Yields vs. land use: How the Green Revolution enabled us to feed a growing population. Retrieved from: https://ourworldindata.org/yields-vs-land-use-how-has-the-world-produced-enough-food-for-a-growing-population (accessed on 17 October 2024).

289. Ritchie, H., Rosado, P., & Roser, M. (2022). Crop yields. Published online at OurWorldInData.org. Retrieved from: https://ourworldindata.org/crop-yields (accessed on 17 October 2024).

290. Robyn, S., & Winter-Nelson, A. (2008). Consumer acceptance of provitamin A-biofortified maize in Maputo, Mozambique. *Food Policy*, *33*, 341–351.

291. Rockström, J., et al. (2009). Planetary boundaries: Exploring the safe operating space for humanity. *Ecology and Society*, *14*, 1–33.

292. Rodríguez, H., & Fraga, R. (1999). Phosphate solubilizing bacteria and their role in plant growth promotion. *Biotechnology Advances*, *17*(4, 5), 319–339.

293. Rodríguez-Eugenio, N., et al. (2018). Soil pollution: A hidden reality. Food and Agriculture Organization of the United Nations, Rome.

294. Rosado, J. L., Hambidge, K. M., Miller, L. V., Garcia, O. P., Westcott, J., Gonzalez, K., et al. (2009). The quantity of zinc absorbed from wheat in adult women is enhanced by biofortification. *Journal of Nutrition*, *139*, 1920–1925.

295. Roy, S., Liu, W., Nandety, R. S., Crook, A., Mysore, K. S., Pislariu, C. I., Frugoli, J., Dickstein, R., & Udvardi, M. K. (2020). Celebrating 20 years of genetic discoveries in legume nodulation and symbiotic nitrogen fixation. *Plant Cell*, *32*(1), 15–41.

296. Ruel, M. T., & Alderman, H. (2013). Nutrition-sensitive interventions and programmes: How can they help to accelerate progress in improving maternal and child nutrition? *Lancet*, *382*(9891), 536–551.

297. Sabaté, J., Harwatt, H., & Soret, S. (2016). Environmental nutrition: A new frontier for public health. *American Journal of Public Health*, *106*(5), 815–821.

298. Saboor, A., Ali, M. A., Danish, S., Ahmed, N., Fahad, S., Datta, R., Ansari, M. J., Nasif, O., Rahman, M. H., & Glick, B. R. (2021). Effect of arbuscular mycorrhizal fungi on the physiological functioning of maize

under zinc-deficient soils. *Scientific Reports*, *11*(1), 1–11.

299. Saeed, Q., Xiukang, W., Haider, F. U., Kučerik, J., Mumtaz, M. Z., Holatko, J., Naseem, M., Kintl, A., Ejaz, M., Naveed, M., et al. (2021). Rhizosphere bacteria in plant growth promotion, biocontrol, and bioremediation of contaminated sites: A comprehensive review of effects and mechanisms. *International Journal of Molecular Sciences*, *22*, 10529.
300. Sahoo, D., Madhu, M., Muralidharan, P., & Sikka, A. (2015). Land management practices for resource conservation under vegetable cultivation in Nilgiris hills ecosystem. *Journal of Environmental Biology*, *36*, 1039–1044.
301. Saia, S., Rappa, V., Ruisi, P., Abenavoli, M. R., Sunseri, F., Giambalvo, D., Frenda, A. S., & Martinelli, F. (2015). Soil inoculation with symbiotic microorganisms promotes plant growth and nutrient transporter genes expression in durum wheat. *Frontiers in Plant Science*, *6*, 815.
302. Salemdeeb, R., Font Vivanco, D., Al-Tabbaa, A., & Zu Ermgassen, E. K. (2017). A holistic approach to the environmental evaluation of food waste prevention. *Waste Management*, *59*, 442–450.
303. Saltzman, A., Birol, B., Bouis, H. E., Boy, E., De Moura, F. F., Islam, Y., & Pfeiffer, W. H. (2013). Biofortification: Progress toward a more nourishing future. *Global Food Security*, *2*, 9–17.
304. Sanchez, P. A. (2002). Soil fertility and hunger in Africa. *Science*, *295*, 2019–2020.
305. Sanket, A., Ghosh, S., Sahoo, R., Nayak, S., & Das, A. (2017). Molecular identification of acidophilic manganese (Mn)-solubilizing bacteria from mining effluents and their application in mineral beneficiation. *Geomicrobiology Journal*, *34*(1), 71–80.
306. Santos, K., Moure, V., Hauer, V., Santos, A. S., Donatti, L., Galvão, C., et al. (2017). Wheat colonization by an *Azospirillum brasilense* ammonium-excreting strain reveals upregulation of nitrogenase and superior plant growth promotion. *Plant and Soil*, *415*(1), 245–255.
307. Seleiman, M. F., & Hafez, E. M. (2021). Optimizing inputs management for sustainable agricultural development. In *Mitigating Environmental Stresses for Agricultural Sustainability in Egypt* (Vol. 25, No. 1, pp. 487–507).
308. Sharma, S. B., Sayyed, R. Z., Trivedi, M. H., & Gobi, T. A. (2013). Phosphate solubilizing microbes: Sustainable approach for managing phosphorus deficiency in agricultural soils. *SpringerPlus*, *2*(1), 1–14.
309. Shi, Z., Zhang, J., Lu, S., Li, Y., & Wang, F. (2020). Arbuscular mycorrhizal fungi improve the performance of sweet sorghum grown in a

Mo-contaminated soil. *Journal of Fungi*, *6*(2), 44.

310. Shi, Z., Zhang, J., Wang, F., Li, K., Yuan, W., & Liu, J. (2018). Arbuscular mycorrhizal inoculation increases molybdenum accumulation but decreases molybdenum toxicity in maize plants grown in polluted soil. *RSC Advances*, *8*(65), 37069–37076.

311. Shittu, A. M., Akelere, D., & Haile, M. (2018). Food price sparks and volatility in local food markets in Nigeria. *ZEF Discussion Papers on Development Policy* No. 248, Center for Development Research, Bonn.

312. Shorrocks, V. M. (1997). The occurrence and correction of boron deficiency. *Plant and Soil*, *193*(1), 121–148.

313. Singh, R., Singh, H., & Raghubanshi, A. S. (2019). Challenges and opportunities for agricultural sustainability in changing climate scenarios: A perspective on Indian agriculture. *Tropical Ecology*, *60*, 167–185.

314. Small, C., & Naumann, T. (2001). The global distribution of human population and recent volcanism. *Global Environmental Change. Part B: Environmental Hazards*, *3*, 93–109.

315. Smil, V. (2002). Nitrogen and food production: Proteins for human diets. *AMBIO: A Journal of the Human Environment*, *31*, 126–131.

316. Smith, J. H., Nayak, D., Albanito, F., Balana, B., Black, H., Boke, S., et al. (2019). Treatment of organic resources before soil incorporation in semi-arid regions improves resilience to El Niño and increases crop production and economic returns. *Environmental Research Letters*, *14*(8), 721–759.

317. Smith, M. J., Shoolery, J., Schwyn, B., Holden, I., & Neilands, J. (1985). Rhizobactin, a structurally novel siderophore from *Rhizobium meliloti*. *Journal of the American Chemical Society*, *107*(6), 1739–1743.

318. Smith, P. (2013). Delivering food security without increasing pressure on land. *Global Food Security*, *2*, 18–23.

319. Snoj Tratnik, J., et al. (2022). Cadmium exposure in adults across Europe: Results from the HBM4EU Aligned Studies survey 2014–2020. *International Journal of Hygiene and Environmental Health*, *246*, 114050.

320. Solomon, R., Wein, T., Levy, B., Eshed, S., Dror, R., Reiss, V., Zehavi, T., Furman, O., Mizrahi, I., & Jami, E. (2022). Protozoa populations are ecosystem engineers that shape prokaryotic community structure and function of the rumen microbial ecosystem. *ISME Journal*, *16*, 1187–1197.

321. Song, Z., Gao, H., Zhu, P., Peng, C., Deng, A., Zheng, C., et al. (2015). Organic amendments increase corn yield by enhancing soil resilience to climate change. *The Crop Journal*, *3*(2), 110–117.

322. Stark, J. (1994). Causes of soil nutrient heterogeneity at different scales. In M. Caldwell & R. Pearcy (Eds.), *Exploitation of Environmental*

Heterogeneity by Plants (pp. 255–284). Academic Press.

323. Steffen, W., Broadgate, W., Deutsch, L., Gaffney, O., & Ludwig, C. (2015). The trajectory of the Anthropocene: The great acceleration. *Anthropocene Review*, *2*, 81–98.
324. Sun, L., Ma, Y., Li, B., Xiao, C., Fan, L., & Xiong, Z. (2018). Nitrogen fertilizer in combination with an ameliorant mitigated yield-scaled greenhouse gas emissions from a coastal saline rice field in southeastern China. *Environmental Science and Pollution Research*, *25*(16), 15896–15908.
325. Svistoonoff, S., Hocher, V., & Gherbi, H. (2014). Actinorhizal root nodule symbioses: What is signaling telling on the origins of nodulation. *Current Opinion in Plant Biology, 20,* 11–18.
326. Swinnen, J., & McDermott, J. (2020). COVID-19 & global food security. International Food Policy Research Institute (IFPRI), Washington, DC.
327. Szilagyi-Zecchin, V. J., Mógor, Á. F., & Figueiredo, G. G. O. (2016). Strategies for characterization of agriculturally important bacteria. In *Microbial Inoculants in Sustainable Agricultural Productivity* (pp. 1–21). Springer.
328. Takahashi, H., Kopriva, S., Giordano, M., Saito, K., & Hell, R. (2011). Sulfur assimilation in photosynthetic organisms: Molecular functions and regulations of transporters and assimilatory enzymes. *Annual Review of Plant Biology, 62,* 157–184.
329. Talsma, E. F., Brouwer, I. D., Verhoef, H., Mbera, G. N., Mwangi, A. M., Demir, A. Y., Maziya-Dixon, B., et al. (2016). Biofortified yellow cassava and vitamin A status of Kenyan children: A randomized controlled trial. *American Journal of Clinical Nutrition, 103*(1), 258–267.
330. Talsma, E. F., Melse-Boonstra, A., de Kok, B. P., Mbera, G. N., Mwangi, A. M., & Brouwer, I. D. (2013). Biofortified cassava with pro-vitamin A is sensory and culturally acceptable for consumption by primary school children in Kenya. *PLOS ONE, 8*(8).
331. Tan, Z. X., Lal, R., & Wiebe, K. D. (2005). Global soil nutrient depletion and yield reduction. *Journal of Sustainable Agriculture, 26,* 123–146.
332. Tilman, D., Balzer, C., Hill, J., & Befort, B. L. (2011). Global food demand and the sustainable intensification of agriculture. *Proceedings of the National Academy of Sciences of the USA, 108,* 20260–20264.
333. Timofeeva, A. M., Galyamova, M. R., & Sedykh, S. E. (2023). Plant growth-promoting soil bacteria: Nitrogen fixation, phosphate solubilization, siderophore production, and other biological activities. *Plants, 12,* 4074.
334. Trotter, P., & Mugisha, M. (2020). Cold chains can help mitigate the

COVID-19 food crisis: Key lessons from Uganda. Oxford Martin School Blog.

335. Tubiello, F., Schmidhuber, J., Howden, M., Neofotis, P. G., Park, S., Fernandes, E., et al. (2008). Climate change response strategies for agriculture: Challenges and opportunities for the 21st century. In *Agriculture Rural Development Discussion Paper* (pp. 42–103).

336. UK Center for Ecology & Hydrology. (2024). Why do soil microbes matter? Retrieved from: https://www.ceh.ac.uk/why-do-soil-microbes-matter (accessed on 17 October 2024).

337. United Nations, Department of Economic and Social Affairs, Population Division. (2019). *World Population Prospects 2019: Highlights* (ST/ESA/SER.A/423). Retrieved from:—https://www.un.org/development/desa/publications/world-population-prospects-2019-highlights.html; https://population.un.org/wpp/publications/files/wpp2019_highlights.pdf

338. United States Department of Agriculture. (2017). U.S. Food Waste Challenge. Retrieved from: https://www.usda.gov/oce/foodwaste/faqs.htm. —https://www.usda.gov/sites/default/files/documents/RD_Food_Loss_and_Waste_Guide.pdf

339. United States Environmental Protection Agency. (2017). *Sustainable Management of Food*. Retrieved from: https://www.epa.gov/sustainable-management-food (accessed on 17 October 2024).

340. United States Environmental Protection Agency. (2024). *How to Prevent Wasted Food Through Source Reduction*. Retrieved from: https://www.epa.gov/sustainable-management-food/how-prevent-wasted-food-through-source-reduction (accessed on 17 October 2024).

341. Upadhyay, S. K., Srivastava, A. K., Rajput, V. D., Chauhan, P. K., Bhojiya, A. A., Jain, D., Chaubey, G., Dwivedi, P., Sharma, B., & Minkina, T. (2022). Root exudates: Mechanistic insight of plant growth-promoting rhizobacteria for sustainable crop production. *Frontiers in Microbiology, 13,* 916488.

342. Urgesa, A. A., Abegaz, A., Bahir, A. L., & Antille, D. L. (2016). Population growth and other factors affecting land-use and land-cover changes in north-eastern Wollega, Ethiopia. *Tropical Agriculture, 93*(4), 298–311.

343. USDA Economic Research Service. (2020). *Agricultural Baseline Database.* Retrieved from: https://www.ers.usda.gov/topics/farm-economy/agricultural-baseline/ (accessed on 17 October 2024).

344. USDA. (2022). *Back to Basics: All About MyPlate Food Groups*. Retrieved from: https://www.usda.gov/media/blog/2017/09/26/back-basics-all-about-myplate-food-groups (accessed on 17 October 2024).

345. Uzoma, K. C., Inoue, M., Andry, H., Fujimaki, H., Zahoor, A., & Nishihara, E. (2011). Effect of cow manure biochar on maize productivity under sandy soil condition. *Soil Use and Management, 27*(2), 205–212.

346. Vaezi, A. R., Zarrinabadi, E., & Auerswald, K. (2017). Interaction of land use, slope gradient, and rain sequence on runoff and soil loss from weakly aggregated semi-arid soils. *Soil & Tillage Research, 172,* 22–31.

347. Vaid, S. K., Kumar, B., Sharma, A., Shukla, A., & Srivastava, P. (2014). Effect of Zn solubilizing bacteria on growth promotion and Zn nutrition of rice. *Journal of Soil Science and Plant Nutrition, 14*(4), 889–910.

348. Van Der Heijden, M. G. A., Bardgett, R. D., & Van Straalen, N. M. (2008). The unseen majority: Soil microbes as drivers of plant diversity and productivity in terrestrial ecosystems. *Ecology Letters, 11,* 296–310.

349. Van Der Heijden, M. G. A., Streitwolf-Engel, R., Riedl, R., Siegrist, S., Neudecker, A., Ineichen, K., Boller, T., Wiemken, A., & Sanders, I. R. (2006). The mycorrhizal contribution to plant productivity, plant nutrition and soil structure in experimental grassland. *New Phytologist, 172,* 739–752.

350. van Jaarsveld, P. J., Faber, M., Tanumihardjo, S. A., Nestel, P., Lombard, C. J., & Benadé, A. J. (2005). β-Carotene-rich orange-fleshed sweet potato improves the vitamin A status of primary school children assessed with the modified-relative-dose-response test. *American Journal of Clinical Nutrition, 81*, 1080–1087.

351. Van Meijl, H., et al. (2017). Challenges of global agriculture in a climate change context by 2050 (AgCLIM50). *JRC Sci. Policy Report*, EUR 28649 EN.

352. Van Velzen, R., Holmer, R., Bu, F., Rutten, L., Van Zeijl, A., Liu, W., et al. (2018). Comparative genomics of the nonlegume *Parasponia* reveals insights into evolution of nitrogen-fixing rhizobium symbioses. *PNAS USA*, *115*(20), E4700–E47E9.

353. Vance, C. (2001). Symbiotic nitrogen fixation and phosphorus acquisition. Plant nutrition in a world of declining renewable resources. *Plant Physiology*, *127*, 390–397.

354. Vermeij, P., Wietek, C., Kahnert, A., Wuèest, T., & Kertesz, M. A. (1999). Genetic organization of sulphur-controlled aryl desulphonation in *Pseudomonas putida* S-313. *Molecular Microbiology*, *32*(5), 913–926.

355. Via, M. (2012). The malnutrition of obesity: Micronutrient deficiencies that promote diabetes. *ISRN Endocrinology*, *2012*, 103472.

356. Vílchez, J. I., Yang, Y., He, D., Zi, H., Peng, L., Lv, S., Kaushal, R., Wang, W., Huang, W., Liu, R., Lang, Z., Miki, D., Tang, K., Paré, P. W., Song,

C. P., Zhu, J. K., & Zhang, H. (2020). DNA demethylases are required for myo-inositol-mediated mutualism between plants and beneficial rhizobacteria. *Nature Plants*, *6*(8), 983–995.

357. Vitousek, P. M., et al. (2009). Nutrient imbalances in agricultural development. *Science*, *324*, 1519–1520.

358. Vitousek, P. M., Porder, S., Houlton, B., & Chadwick, O. (2010). Terrestrial phosphorus limitation: Mechanisms, implications, and nitrogen–phosphorus interactions. *Ecological Applications*, *20*, 5–15.

359. Vives-Peris, V., de Ollas, C., Gómez-Cadenas, A., & Pérez-Clemente, R. M. (2020). Root exudates: From plant to rhizosphere and beyond. *Plant Cell Reports*, *39*, 3–17.

360. Vorholt, J. A., Vogel, C., Carlström, C. I., & Müller, D. B. (2017). Establishing causality: Opportunities of synthetic communities for plant microbiome research. *Cell Host Microbe*, *22*, 142–155.

361. Wagg, C., Bender, S. F., Widmer, F., & van der Heijden, M. G. A. (2014). Soil biodiversity and soil community composition determine ecosystem multifunctionality. *Proceedings of the National Academy of Sciences USA*, *111*, 5266–5270.

362. Wahab, A., Muhammad, M., Munir, A., Abdi, G., Zaman, W., Ayaz, A., Khizar, C., & Reddy, S. P. P. (2023). Role of arbuscular mycorrhizal fungi in regulating growth, enhancing productivity, and potentially influencing ecosystems under abiotic and biotic stresses. *Plants*, *12*, 3102.

363. Walker, T. W., & Syers, J. K. (1976). The fate of phosphorus during pedogenesis. *Geoderma*, *15*, 1–19.

364. Wang, P., Alpert, P., & Yu, F. H. (2016). Clonal integration increases relative competitive ability in an invasive aquatic plant. *American Journal of Botany*, *103*, 2079–2086.

365. Wang, S., Chen, A., Xie, K., Yang, X., Luo, Z., Chen, J., Zeng, D., Ren, Y., Yang, C., Wang, L., Feng, H., López-Arredondo, D. L., Herrera-Estrella, L. R., & Xu, G. (2020). Functional analysis of the OsNPF4.5 nitrate transporter reveals a conserved mycorrhizal pathway of nitrogen acquisition in plants. *PNAS USA*, *117*(28), 16649–16659.

366. Wang, X., He, T., Gen, S., Zhang, X. Q., Wang, X., Jiang, D., Li, C., Li, C., Wang, J., Zhang, W., et al. (2020). Soil properties and agricultural practices shape microbial communities in flooded and rainfed croplands. *Applied Soil Ecology*, *147*, 103449.

367. Wani, A. K., Akhtar, N., Sher, F., Navarrete, A. A., & Américo-Pinheiro, J. H. P. (2022). Microbial adaptation to different environmental conditions: Molecular perspective of evolved genetic and cellular systems. *Archives*

of Microbiology, *204*, 144.

368. Watts-Williams, S. J., & Cavagnaro, T. R. (2018). Arbuscular mycorrhizal fungi increase grain zinc concentration and modify the expression of root ZIP transporter genes in a modern barley (*Hordeum vulgare*) cultivar. *Plant Science*, *274*, 163–170.
369. Weisskopf, L., Schulz, S., & Garbeva, P. (2021). Microbial volatile organic compounds in intra-kingdom and inter-kingdom interactions. *Nature Reviews Microbiology*, *19*(6), 391–404.
370. WFP. (2020b). Global monitoring of school meals during COVID-19 school closures. *World Food Programme (WFP)*, Rome, 15 October 2020.
371. Whitelaw, M., Harden, T., & Helyar, K. (1999). Phosphate solubilisation in solution culture by the soil fungus *Penicillium radicum*. *Soil Biology and Biochemistry*, *31*(5), 655–665.
372. WHO. (2024). *World Health Organization Global Hunger Index*. Available from: https://www.who.int/data/nutrition/nlis/info/global-hunger-index-(ghi) (accessed on 17 October 2024).
373. Wright, C. L., & Lehtovirta-Morley, L. E. (2023). Nitrification and beyond: Metabolic versatility of ammonia oxidising archaea. *ISME Journal*, *17*, 1358–1368.
374. Xiong, W., Song, Y., Yang, K., Gu, Y., Wei, Z., Kowalchuk, G. A., Xu, Y., Jousset, A., Shen, Q., & Geisen, S. (2020). Rhizosphere protists are key determinants of plant health. *Microbiome*, *8*, 27.
375. Xu, G. H., Chague, V., Melamed-Bessudo, C., Kapulnik, Y., Jain, A., Raghothama, K. G., Levy, A. A., & Silber, A. (2007). Functional characterization of LePT4: A phosphate transporter in tomato with mycorrhiza-enhanced expression. *Journal of Experimental Botany*, *58*(10), 2491–2501.
376. Xue, W., Huang, L., Yu, F. H., & Bezemer, T. M. (2018). Intraspecific aggregation and soil heterogeneity: Competitive interactions of two clonal plants with contrasting spatial architecture. *Plant and Soil*, *425*, 231–240.
377. Yi, W., Ziyu, Z., Shujun, Y., Hao, L., Noman, A., & Zhang, S. H. (2022). Cellulose degradation microorganisms and environmental-friendly solution to agricultural waste management. In R. Prasad & S. H. Zhang (Eds.), *Beneficial Microorganisms in Agriculture* (pp. 307–328). Springer Nature.
378. Zamioudis, C., Korteland, J., Van Pelt, J. A., Van Hamersveld, M., Dombrowski, N., Bai, Y., et al. (2015). Rhizobacterial volatiles and photosynthesis-related signals coordinate MYB 72 expression in *Arabidopsis* roots during onset of induced systemic resistance and iron-

deficiency responses. *Plant Journal*, *84*(2), 309–322.

379. Zeder, M. A. (2011). The origins of agriculture in the Near East. *Current Anthropology*, *52*, S221–S225.

380. Zerssa, G. W., Kim, D. G., Koal, P., & Eichler-Löbermann, B. (2021). Combination of compost and mineral fertilizers as an option for enhancing maize (*Zea mays* L.) yields and mitigating greenhouse gas emissions from a Nitisol in Ethiopia. *Agronomy*, *11*(11), 2097.

381. Zewdu, E., Hadgu, G., & Nigatu, L. (2020). Impacts of climate change on sorghum production in North Eastern Ethiopia. *African Journal of Environmental Science and Technology*, *14*(2), 49–63.

382. Zhang, F., Du, P., Song, C. X., & Wu, Q. S. (2015). Alleviation of mycorrhiza to magnesium deficiency in trifoliate orange: Changes in physiological activity. *Emirates Journal of Food and Agriculture*, *27*, 763–769.

383. Zhang, H., Sun, Y., Xie, X., Kim, M. S., Dowd, S. E., & Paré, P. W. (2009). A soil bacterium regulates plant acquisition of iron via deficiency-inducible mechanisms. *Plant Journal*, *58*(4), 568–577.

384. Zhang, H., Xie, X., Kim, M. S., Kornyeyev, D. A., Holaday, S., & Paré, P. W. (2008). Soil bacteria augment *Arabidopsis* photosynthesis by decreasing glucose sensing and abscisic acid levels in planta. *Plant Journal*, *56*(2), 264–273.

385. Zhang, J., Xing, P., Niu, M., Wei, G., & Shi, P. (2021). Taxonomic compositions and co-occurrence relationships of protists in bulk soil and rhizosphere of soybean fields in different regions of China. *Frontiers in Microbiology*, *12*, 738129.

386. Zhang, L. M., Alpert, P., & Yu, F. H. (2022). Nutrient foraging ability promotes intraspecific competitiveness in the clonal plant *Hydrocotyle vulgaris*. *Ecological Indicators*, *138*, 108862.

387. Zhao, F., Lehmann, J., Solomon, D., Fox, M., & McGrath, S. (2006). Sulphur speciation and turnover in soils: Evidence from sulphur K-edge XANES spectroscopy and isotope dilution studies. *Soil Biology and Biochemistry*, *38*(5), 1000–1007.

388. Zhou, C., Guo, J., Zhu, L., Xiao, X., Xie, Y., Zhu, J., et al. (2016). *Paenibacillus polymyxa* BFKC01 enhances plant iron absorption via improved root systems and activated iron acquisition mechanisms. *Plant Physiology and Biochemistry*, *105*, 162–173.

389. Zhou, J., Dong, B. C., Alpert, P., Li, H. L., Zhang, M. X., Lei, G. C., & Yu, F. H. (2012). Effects of soil nutrient heterogeneity on intraspecific competition in the invasive, clonal plant *Alternanthera philoxeroides*. *Annals of Botany*, *109*, 813–818.

390. Zhu, C., Delmont, T. O., Vogel, T. M., & Bromberg, Y. (2015). Functional basis of microorganism classification. *PLoS Computational Biology*, *11*, e1004472.

INDEX

FSC
www.fsc.org
MIX
Papier | Fördert gute Waldnutzung
FSC® C083411

Kolibri 360 GmbH
Ferdinand-Jühlke-Straße 7
99095 Erfurt, Deutschland
produktsicherheit@kolibri360.de